Thomas Hell

Modeling the Thermodynamics of QCD

Thomas Hell

Modeling the Thermodynamics of QCD

Using Nambu and Jona-Lasinio-type Covariant Nonlocal Chiral Quark Models Extended by the Polyakov Loop

Südwestdeutscher Verlag für Hochschulschriften

Impressum/Imprint (nur für Deutschland/ only for Germany)

Bibliografische Information der Deutschen Nationalbibliothek: Die Deutsche Nationalbibliothek verzeichnet diese Publikation in der Deutschen Nationalbibliografie; detaillierte bibliografische Daten sind im Internet über http://dnb.d-nb.de abrufbar.

Alle in diesem Buch genannten Marken und Produktnamen unterliegen warenzeichen-, marken- oder patentrechtlichem Schutz bzw. sind Warenzeichen oder eingetragene Warenzeichen der jeweiligen Inhaber. Die Wiedergabe von Marken, Produktnamen, Gebrauchsnamen, Handelsnamen, Warenbezeichnungen u.s.w. in diesem Werk berechtigt auch ohne besondere Kennzeichnung nicht zu der Annahme, dass solche Namen im Sinne der Warenzeichen- und Markenschutzgesetzgebung als frei zu betrachten wären und daher von jedermann benutzt werden dürften.

Verlag: Südwestdeutscher Verlag für Hochschulschriften GmbH & Co. KG
Dudweiler Landstr. 99, 66123 Saarbrücken, Deutschland
Telefon +49 681 37 20 271-1, Telefax +49 681 37 20 271-0
Email: info@svh-verlag.de
Zugl.: München, TU, Diss., 2010

Herstellung in Deutschland:
Schaltungsdienst Lange o.H.G., Berlin
Books on Demand GmbH, Norderstedt
Reha GmbH, Saarbrücken
Amazon Distribution GmbH, Leipzig
ISBN: 978-3-8381-2046-1

Imprint (only for USA, GB)

Bibliographic information published by the Deutsche Nationalbibliothek: The Deutsche Nationalbibliothek lists this publication in the Deutsche Nationalbibliografie; detailed bibliographic data are available in the Internet at http://dnb.d-nb.de.

Any brand names and product names mentioned in this book are subject to trademark, brand or patent protection and are trademarks or registered trademarks of their respective holders. The use of brand names, product names, common names, trade names, product descriptions etc. even without a particular marking in this works is in no way to be construed to mean that such names may be regarded as unrestricted in respect of trademark and brand protection legislation and could thus be used by anyone.

Publisher: Südwestdeutscher Verlag für Hochschulschriften GmbH & Co. KG
Dudweiler Landstr. 99, 66123 Saarbrücken, Germany
Phone +49 681 37 20 271-1, Fax +49 681 37 20 271-0
Email: info@svh-verlag.de

Printed in the U.S.A.
Printed in the U.K. by (see last page)
ISBN: 978-3-8381-2046-1

Contents

Contents

1 Introduction

Quantum chromodynamics (QCD)—the theory of the strong interaction of quarks and gluons—is characterized by two important features: the first one is asymptotic freedom, the weakening of the interaction strength at small distances or, equivalently, high energies; the second one is confinement, i. e., the phenomenon that no free quarks exist in nature. Given these properties, it is natural to assume that QCD matter at high energy densities undergoes a phase transition from a state with confined hadrons into a new state of matter with liberated quarks and gluons.

It is believed that quantum chromodynamics at finite temperature, T, and finite baryon density, ϱ_B, has a rich phase structure governed by symmetries and their (spontaneous) breakdown. Besides being invariant under the local $SU(3)_c$ color gauge group, the QCD Lagrangian possesses a $\mathcal{G} := SU(3)_L \times SU(3)_R \times U(1)_L \times U(1)_R$ global chiral symmetry, when only three massless flavors "up", "down" and "strange" are considered. The $U(1)_A$ axial symmetry is only conserved at a classical level, quantum corrections break it down to the center symmetry $Z(6)$. At low temperature T and low baryon density ϱ_B, $\mathcal{G}$ is broken down spontaneosly to $\mathcal{G} \to SU(3)_V \times U(1)_B$. The breakdwon of $\mathcal{G}$ in this region is characterized by a nonvanishing chiral condensate $\langle \bar{\psi}\psi \rangle \neq 0$ which is the corresponding order parameter. According to Goldstone's theorem, massless pseudoscalar bosons are associated with the spontaneous chiral symmetry breaking. These are identified with the pions, kaons and the eta meson. At temperatures typically around the intrinsic scale of QCD, i. e., $T \sim \Lambda_{QCD} \sim 200\,\mathrm{MeV}$ the chiral condensates melt away and a quark-gluon plasma symmetric under $SU(3)_c$ and $SU(3)_L \times SU(3)_R$ is formed, with chiral symmetry restored in its trivial (Wigner-Weyl) realization. The experimental exploration of the quark-gluon plasma is pursued using ultrarelativistic heavy-ion collisions at the Relativistic Heavy Ion Collider (RHIC) at Brookhaven National Laboratory (BNL) and, in the future, at the Large Hadron Collider (LHC) at CERN in Geneva. Finally, in cold quark matter at low temperature T and high densities $\varrho_B \sim \Lambda^3_{QCD} \sim 1\,\mathrm{fm}^{-3}$ condensation of quark-quark pairs leads to a color superconducting phase and dynamical breaking of color $SU(3)_c$ gauge symmetry. This is exactly what one would expect from ordinary BCS-superconductors: all gluons acquire masses, as required by the Anderson-Higgs mechanism in order to describe the Meissner effect. There are many different color-superconducting phases. At asymptotically high densities, the so-called color-flavor-locked phase is conjectured, characterized by the symmetry breaking $\mathcal{G} \to SU(3)_{c+V} \times Z(2)$. There are speculations that the interior of compact stellar objects such as neutron stars could be formed by such color-superconducting matter.

In this work we are going to investigate the phase structure of strongly interacting matter limiting ourself to the chiral and confinement-deconfinement transitions, i. e., we are not considering color superconductors. On the theoretical side there are basically two different strategies to gather further insight into the phase structure of strongly interacting matter: large-scale computer simulations of full QCD on a discretized lattice, and model calculations based on fundamental symmetries of QCD. In this work we follow the second route and choose as starting point a model proposed long ago by Y. Nambu and G. Jona-Lasinio, the NJL model.

Models of the Nambu and Jona-Lasinio type have been quite useful for orientation in this context as they properly incorporate the chiral symmetry breaking scenario of low-energy QCD. A basic element of such models is the gap equation connecting the chiral condensate and the dynamical quark mass, providing a mechanism for spontaneous chiral symmetry breaking and the generation of quark quasiparticle masses. Thermodynamic aspects of confinement, while absent in the original NJL model, can be implemented by a synthesis with Polyakov-loop dynamics. The resulting PNJL model has been remarkably successful in describing the two-flavor thermodynamics of QCD.

The original PNJL approach worked with an artificial momentum-space cutoff, $\Lambda_{\mathrm{NJL}} \approx$ $\approx (0.6\text{–}0.7)\,\mathrm{GeV}$, which prohibits establishing connections with well-known properties of QCD at higher momentum scales such as the running coupling and the momentum-dependent quark-mass function. Furthermore, a meaningful extrapolation to the high-density region with its variety of color-superconducting phases cannot be performed once the quark Fermi momentum becomes comparable to the NJL cutoff.

In this work we are going to derive a nonlocal extension of the original PNJL model that does not have such *a priori* limitations. We are treating the two-flavor case first and then extend to the more realistic three-flavor case. These nonlocal PNJL models remove the problem of the cutoff regularization by introducing momentum-dependent quark interactions that permit to realize the high-momentum interface with QCD and Dyson-Schwinger calculations at the level of the quark quasiparticle propagators. The extension to the three-flavor case, including the strange quark, is nontrivial. This step involves a detailed study of the axial U(1) anomaly, its role in separating the flavor singlet component of the pseudoscalar meson nonet from the Nambu-Goldstone boson sector, and its thermodynamical implications.

This work is organized as follows. In the first part (Chapts. 2 and 3) we deal with zero temperatures and densities. The second part (Chapts. 4 and 5) is devoted to the thermodynamics of strongly interacting matter. In Chapt. 2 we introduce the basic elements of quantum chromodynamics and describe, in particular, two approaches to nonperturbative QCD: lattice calculations and Dyson-Schwinger equations. Chapt. 3 describes in detail the derivation of the nonlocal Nambu–Jona-Lasinio model, starting from a color-current interaction, as it arises in QCD. First, we consider the case of two flavors and present the formalism necessary for the derivation of the fundamental gap equation. Meson properties will be calculated, and we derive low-energy theorems known from current-algebra. The second part of Chapt. 3 is devoted to a detailed study of the axial $U(1)_A$ anomaly. We refer to the instanton-liquid model and show how this is related to the anomalous breaking of the $U(1)_A$ symmetry and the large mass of the eta-prime meson, $m_{\eta'} \simeq 958\,\mathrm{MeV}$, about 400 MeV higher than the eta mass. In the interpretation given by 't Hooft, the axial anomaly is induced by instanton effects and translates into an axial $U(1)_A$ breaking effective interaction between quarks that has the form of an $N_\mathrm{f} \times N_\mathrm{f}$

determinant of right- and left-handed quark bilinears, $\psi_i(1 \pm \gamma_5)\psi_j$. For $N_{\mathrm{f}} = 3$ this is a genuine six-point vertex involving all three up, down and strange quarks simultaneously. The instanton-liquid model provides us with the nonlocal extension of this genuine six-fermion term. Given this expression, we can apply the formalism known from the two-flavor case to the three-flavor version, as well, ending with formulas for the pseudoscalar mesons, their decay constants and fundamental low-energy theorems. In Chapt. 4 we give an overview of the present wisdom about the QCD phase diagram. We describe the experimental and theoretical status. Because of the recent developments in finite-temperature lattice calculations, we devote an extra section to this issue. Chapt. 5 describes the thermodynamics of the nonlocal model developed in this work. First, we prove why the Polyakov loop serves as the order parameter for the confinement-deconfinement transition. Next, we couple the Polyakov loop to the nonlocal NJL model, leading to the nonlocal Polyakov-loop-extended NJL (PNJL) model. Using this model, we describe the finite-temperature and -density region of the QCD phase diagram, for the both the two- and three-flavor case. Before presenting our conclusions in Chapt. 7, we describe in Chapt. 6 a formal derivation of the nonlocal PNJL model starting directly from full QCD. This is done consistently for the three-flavor case. This derivation justifies rigorously the assumptions made in the nonlocal PNJL model, and it sheds much light on its physical content. The Appendices collect the notations used in the present work and outline details of some calculations.

2 Quantum Chromodynamics

Quantum chromodynamics (QCD) is the theory of the strong interaction. QCD is a non-Abelian gauge theory with gauge group SU(3). The gauge bosons (gluons) couple to fermions (quarks) in the fundamental triplet representation [**3**] of the gauge group. The gluons are in the adjoint octet representation [**8**] of SU(3).

After a short historical overview we introduce the QCD Lagrangian and discuss the symmetries and symmetry breaking pattern investigated in this work. We explain the physical and mathematical tools that will be used throughout the work. Following a brief description of lattice QCD techniques we present results obtained from the Dyson-Schwinger formalism that will be, although in a slightly modified form, the framework applied here.

2.1 The Eightfold Way

At the beginning of the 1960's Murray Gell-Mann developed a theory collecting baryons and mesons into (flavor-) SU(3) octets; this is the well-known *Eightfold Way*. In 1962, Gell-Mann applied this scheme also to the spin-$\frac{3}{2}$ decuplet states and predicted a state with electrical charge -1, spin $\frac{3}{2}$ and strangeness -3, that he called Ω^-. Its discovery two years later at Brookhaven National Laboratory (BNL) became a triumphant success of the Eightfold Way. On the other hand, such a (fermionic) state should not exist as a consequence of Pauli's exclusion principle. Assuming that the Ω^- is composed of three (indistinguishable) strange quarks, a state with spin $\frac{3}{2}$ and, for example, spin projection $M_s = +\frac{3}{2}$ has all three quark spins aligned in the same direction. Hence the spin-flavor wave function of this state is totally symmetric, in contradiction to Pauli's principle which dictates that the wave function of a fermion system must be totally antisymmetric. To reconcile the baryon spectrum with the spin-statistics theorem, Han and Nambu (1965), Greenberg (1965), and Gell-Mann (1972) proposed that quarks carry an additional quantum number, later called *color*[1]. Furthermore, they introduced the *ad hoc* assumption that baryon wave functions must be totally antisymmetric in color quantum numbers. If this is the case, then quark wave functions that are totally symmetric in spin and flavor are totally antisymmetric overall.

Independently of Gell-Mann's studies, the Isreali physicist Yuval Ne'eman proposed a similar theory and initiated the subsequent development of the quark model, a more general classification scheme for hadrons on the basis of their valence quarks. All quarks are assigned a baryon number of $B = \frac{1}{3}$. The up, charm and top quarks (u, c, t respectively) have an electric charge of $Q = +\frac{2}{3}$, while the down, strange and bottom quarks (d, s, b) have an electric charge of

---

[1]In this work we use the three colors *red* (r), *green* (g) and *blue* (b).

$Q = -\frac{1}{3}$. Furthermore, for each quark there exists an antiquark which is described by the fundamental representation $[\overline{3}]$ of SU(3) and is attributed opposite quantum numbers compared to the respective quark. Assuming that up and down quarks have nearly the same mass and that the strange quark is only a little heavier than u and d quarks, it makes sense to consider them altogether as representing an (approximate) SU(3) flavor symmetry.[2] Besides baryon number B, it is useful to collect u and d quarks in an $I = \frac{1}{2}$ isospin doublet with $I_3 = +\frac{1}{2}$ for up and $I_3 = -\frac{1}{2}$ for down, and to attribute the strange quark the strangeness $S = -1$. From this SU(3) quark model follows the so-called Gell-Mann–Nishijima formula[3]

$$Q = I_3 + \frac{1}{2}\left(B + S\right).$$

The SU(3) flavor symmetry structure implies, e. g.,

$$[3] \otimes [\overline{3}] = [8] \oplus [1]$$

for combinations of quarks and antiquarks such as the ones forming the known pseudoscalar meson octet of pions, kaons and eta (η_8), and the singlet eta meson (η_0). The η_0 and η_8 mix in order to form the physical mass eigenstates η and η' observed in experiment.[4] For baryons with their three valence quarks one finds

$$[3] \otimes [3] \otimes [3] = [10] \oplus [8] \oplus [8] \oplus [1].$$

The multiplets on the right-hand side include the spin-$\frac{3}{2}$ baryon decuplet and the spin-$\frac{1}{2}$ baryon octet (comprising the nucleons). The two octets have mixed symmetry; in order to obey Pauli's principle, they have to be recast in symmetric and antisymmetric multiplets.[5] The octet of lowest mass has the proton and neutron as its members. Finally, the singlet state is totally antisymmetric; Pauli's exclusion principle prohibits such a state to exist (in a ground state with zero orbital angular momentum). The two aforementioned combinations are thus the only combinations of quark-antiquark and three quarks[6] forming hadrons (mesons and baryons) that are singlets under color SU(3) transformations.

The introduction of color as an additional degree of freedom created the obvious question of the mechanism behind *confinement*, ensuring that all hadron wave functions are color singlets. The confining phase is usually described in terms of the Polyakov loop, to be discussed in Chapt. 5.2. Furthermore, QCD enjoys the property of asymptotic freedom (see Sect. 2.3.2), which means that the QCD coupling strength decreases when going to higher energies, or equivalently, to smaller distances (as opposed to quantum electrodynamics).

---

[2]The historical development of the quark model started with the three lightest quarks; generalizations to more quarks followed later.

[3]This formula was originally given by Kazuhiko Nishijima and Tadao Nakano in 1953 based on empirical observations.

[4]We will return to this issue and, in particular, to the large mass difference between the η and the η' in Sect. 3.3.1.

[5]This is the reason why only one octet is observed.

[6]Together with $[\overline{3}] \otimes [\overline{3}] \otimes [\overline{3}]$, where quarks are replaced by antiquarks.

2.2 QCD Lagrangian and Symmetries

After this historical overview we can now write down the QCD Lagrangian. As mentioned before, QCD relies on (local) SU(3) color gauge invariance:

$$\mathscr{L}_{\text{QCD}} = \sum_{f \in \text{flavors}} \bar{\psi}_f \left[i\gamma^\mu D_\mu - m_f \right] \psi_f - \frac{1}{2}\text{Tr}(G_{\mu\nu}G^{\mu\nu}), \qquad (2.2.1)$$

where $\psi_f(x) = (\psi_{r,f}(x), \psi_{b,f}(x), \psi_{g,f}(x))^\top$ is the quark field of flavor f, m_f is the current quark mass of mass with flavor f, g is the coupling strength of the strong interaction and the γ^μ are the Dirac matrices. *Local* gauge invariance implies replacing the partial derivative ∂_μ by the *covariant* derivative,

$$D_\mu := \partial_\mu - \mathrm{i}gA_\mu,$$

that incorporates the eight *gluon fields* A_μ^a, $a \in \{1, \dots, 8\}$, where we have defined $A_\mu := t_a A_\mu^a$ using the generators $t_a := \frac{\lambda_a}{2}$ of the Lie algebra of SU(3)$_c$ (in the *fundamental representation*) which satisfy the commutation relations

$$[t_a, t_b] = \mathrm{i}f_{abc}, \qquad \text{Tr}(t_a t_b) = \frac{1}{2}\delta_{ab}, \qquad \text{Tr}(t_a) = 0.$$

The λ_a are the Gell-Mann matrices and the f_{abc} are the totally antisymmetric structure constants. The last term in Eq. (2.2.1) describes the gluon dynamics through the *gluonic field strength tensor*

$$G^{\mu\nu} := \frac{\mathrm{i}}{g}[D^\mu, D^\nu]. \qquad (2.2.2)$$

The Lagrangian $\mathscr{L}_{\text{QCD}}$ is indeed invariant under gauge transformations $U(x) = \exp(-\mathrm{i}t_a\theta^a(x)) \in$ SU(3)$_c$. The quark fields $\psi_f(x)$ are functions of space-time in the fundamental representation of the group and transforms as

$$\psi_f(x) \to U(x)\psi_f(x). \qquad (2.2.3)$$

The gluon fields A_μ are functions in the (inhomogeneous) adjoint representation of SU(3)$_c$ and transform according to

$$A_\mu \to A_{U,\mu} := U\left[A_\mu - \frac{\mathrm{i}}{g}U^{-1}\partial_\mu U\right]U^{-1}. \qquad (2.2.4)$$

As we will discuss in greater detail later (see Sect. 3.3.1), the QCD Lagrangian could additionally include a term

$$\mathscr{L}_\theta = -\frac{\theta}{64\pi^2}\varepsilon^{\mu\nu\rho\sigma}G_{\mu\nu}G_{\rho\sigma}, \qquad (2.2.5)$$

where $\varepsilon_{\mu\nu\rho\sigma}$ is the totally antisymmetric Levi-Civita tensor. The inclusion of this term would violate P and CP conservation. Therefore, a nonzero[7] θ would generate a neutron electric dipole moment d_n proportional to $|\theta|$ (cf. Refs. [1,2]). From the current best upper limit of the dipole moment, $|d_\text{n}| < 2.9 \cdot 10^{-26}\,e \cdot \text{cm}$ [3], one gets $|\theta| < 10^{-10}$. This explains why it is justified to omit $\mathscr{L}_\theta$ from the QCD Lagrangian for our purposes. It is still puzzling, though, why θ is so

---

[7]From the instanton model (Sect. 3.3.1) it follows that, if at least one of the quarks were massless, the theta-term would vanish.

small. One possible solution was given by R. Peccei and H. Quinn [4,5], F. Wilczek [6] and S. Weinberg [7]. The idea is to interpret θ as the Nambu-Goldstone boson ("axion") of a new spontaneously broken U(1) symmetry, the Peccei-Quinn symmetry. Owing to instanton effects in the QCD vacuum, the Peccei-Quinn symmetry is explicitly broken providing a small mass for the axion.

Besides the invariance under the $\mathrm{SU}(3)_\mathrm{c}$ color gauge group, the QCD Lagrangian $\mathscr{L}_{\mathrm{QCD}}$ is Poincaré and CPT invariant and it exhibits other global symmetries. The breaking patterns of these symmetries are summarized in the following diagram for N_f quark flavors:

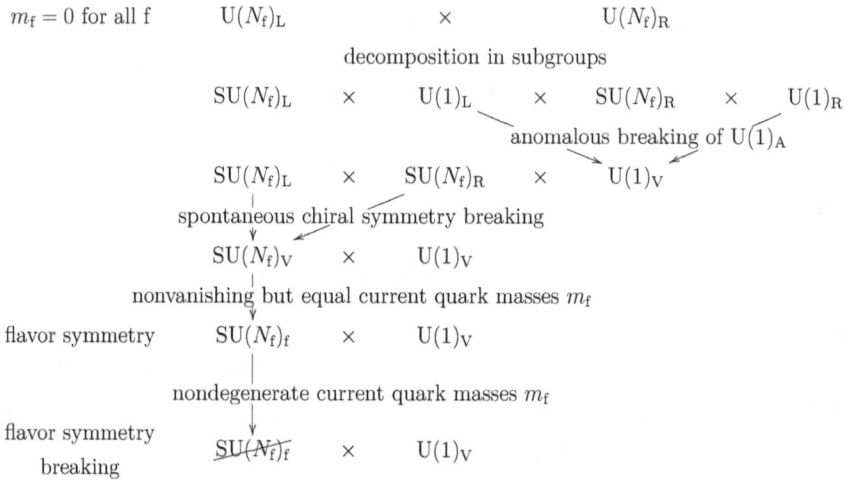

$m_\mathrm{f} = 0$ for all f$\qquad$ $\mathrm{U}(N_\mathrm{f})_\mathrm{L}$ $\qquad\qquad\times\qquad\qquad$ $\mathrm{U}(N_\mathrm{f})_\mathrm{R}$

decomposition in subgroups

$\mathrm{SU}(N_\mathrm{f})_\mathrm{L}$ $\quad\times\quad$ $\mathrm{U}(1)_\mathrm{L}$ $\quad\times\quad$ $\mathrm{SU}(N_\mathrm{f})_\mathrm{R}$ $\quad\times\quad$ $\mathrm{U}(1)_\mathrm{R}$

anomalous breaking of $\mathrm{U}(1)_\mathrm{A}$

$\mathrm{SU}(N_\mathrm{f})_\mathrm{L}$ $\quad\times\quad$ $\mathrm{SU}(N_\mathrm{f})_\mathrm{R}$ $\quad\times\quad$ $\mathrm{U}(1)_\mathrm{V}$

spontaneous chiral symmetry breaking

$\mathrm{SU}(N_\mathrm{f})_\mathrm{V}$ $\quad\times\quad$ $\mathrm{U}(1)_\mathrm{V}$

nonvanishing but equal current quark masses m_f

flavor symmetry$\qquad$ $\mathrm{SU}(N_\mathrm{f})_\mathrm{f}$ $\quad\times\quad$ $\mathrm{U}(1)_\mathrm{V}$

nondegenerate current quark masses m_f

flavor symmetry
breaking $\qquad\qquad$ $\mathrm{SU}(N_\mathrm{f})_\mathrm{f}$ $\quad\times\quad$ $\mathrm{U}(1)_\mathrm{V}$

Let us consider vanishing current quark masses first. In this case, QCD does not discriminate between different flavors. First of all, one observes that $\mathscr{L}_{\mathrm{QCD}}$, Eq. (2.2.1), is invariant under a chiral $\mathrm{U}(N_\mathrm{f})_\mathrm{L} \times \mathrm{U}(N_\mathrm{f})_\mathrm{R}$ flavor symmetry group that can be divided in special unitary $\mathrm{SU}(N_\mathrm{f})$ and unitary $\mathrm{U}(1)$ groups. The chiral $\mathrm{SU}(N_\mathrm{f})_\mathrm{L} \times \mathrm{SU}(N_\mathrm{f})_\mathrm{R}$ symmetry is spontaneously broken by the nontrivial QCD vacuum to the vector $\mathrm{SU}(N_\mathrm{f})_\mathrm{V}$ with the formation of chiral condensates of quark flavors with current quark masses smaller than the typical QCD scale $\Lambda_{\mathrm{QCD}} \approx 250\,\mathrm{MeV}$ (see Sect. 2.4.2). The axial symmetry $\mathrm{U}(1)_\mathrm{A}$ is exact in the classical theory, but broken in the quantum theory by the axial anomaly. Its appearance is strongly related to the nontrivial structure of the QCD vacuum (see Sect. 3.3.1).[8] The vector symmetry, $\mathrm{U}(1)_\mathrm{V}$ corresponds to the baryon number which is an exact symmetry.

For nonvanishing current quark masses, $\mathscr{L}_{\mathrm{QCD}}$ is not invariant under the chiral symmetry $\mathrm{SU}(N_\mathrm{f})_\mathrm{L} \times \mathrm{SU}(N_\mathrm{f})_\mathrm{R}$ because of the mass term $\delta\mathscr{L}_m = \bar{\psi}\hat{m}\psi$, (with the current quark mass matrix $\hat{m} = \mathrm{diag}(m_1, \ldots, m_{N_\mathrm{f}})$) that mixes left- and right-handed components of the Dirac field. However, if all masses are assumed to be equal but nonzero, then the remnant $\mathrm{SU}(N_\mathrm{f})_\mathrm{V}$ symmetry is still a symmetry of $\mathscr{L}_{\mathrm{QCD}}$, while axial $\mathrm{SU}(N_\mathrm{f})_\mathrm{A}$ is explicitly broken. This can easily be seen

---

[8]There is a second symmetry that is anomalously broken: scale, or the so-called conformal symmetry. This conformal anomaly gives rise to asymptotic freedom.

by introducing

$$\tilde{m} := \hat{m} - \frac{1}{N_{\mathrm{f}}}(m_1 + \ldots + m_{N_{\mathrm{f}}}) \cdot 1_{N_{\mathrm{f}} \times N_{\mathrm{f}}}.$$

$\tilde{m}$ is traceless, therefore it can be expressed in terms of elements of the SU(N_{f}) Cartan subalgebra $\mathcal{H}$ (i. e., the subset of SU(N_{f}) matrices that commute with *all* elements of SU(N_{f})). A basis of $\mathcal{H}$ is given by the $N_{\mathrm{f}} \times N_{\mathrm{f}}$ matrices $h_i := \mathrm{diag}(0, \ldots, \underset{i}{1}, \underset{i+1}{-1}, 0, \ldots, 0)$ for $i \in \{1, \ldots, N_{\mathrm{f}} - 1\}$.
Thus, we may write

$$\hat{m} = \frac{1}{N_{\mathrm{f}}}(m_1 + \ldots + m_{N_{\mathrm{f}}}) \cdot 1_{N_{\mathrm{f}} \times N_{\mathrm{f}}} + c_1 h_1 + \ldots + c_{N_{\mathrm{f}}} h_{N_{\mathrm{f}}},$$

with the expansion coefficients $c_i = \frac{1}{2}\mathrm{Tr}\,(\tilde{m} \cdot h_i) = \frac{1}{2}(m_i - m_{i+1}), i \in \{1, \ldots, N_{\mathrm{f}} - 1\}$. From this it is clear, that in the case of isospin symmetry with all current quarks equal, *all* coefficients c_i vanish. In this case, only the term proportional to unity survives. Thus, $\delta\mathcal{L}_m$ is invariant under SU(N_{f})$_{\mathrm{V}}$ transformations of ψ for equal current quark masses, but not invariant under SU(N_{f})$_{\mathrm{A}}$ transformations. The remnant SU(N_{f})$_{\mathrm{V}}$ symmetry is often called flavor symmetry and denoted by SU(N_{f})$_{\mathrm{f}}$. For instance, for $N_{\mathrm{f}} = 3$ one recovers degenerate meson octets as predicted by the eightfold way even before QCD was established. Finally, if the current quark masses are different, then SU(N_{f})$_{\mathrm{f}}$ is explicitly broken and one obtains nondegenerate meson $N_{\mathrm{f}}^2 - 1$ multiplets.

2.3 Perturbative Quantum Chromodynamics

2.3.1 Path-Integral Formalism

Owing to asymptotic freedom, a perturbative approach can be applied to QCD in the high-energy region where the coupling constant becomes weak. One approach to a perturbative analysis of QCD is Feynman's path-integral formalism [8–10]. Consider connected n-point functions $\mathcal{G}^{(n)}$, defined through the vacuum expectation value

$$\mathcal{G}^{(n)} := \langle 0 | \mathcal{T}[\hat{\phi}(x_1) \ldots \hat{\phi}(x_n)] | 0 \rangle = \frac{\int \mathscr{D}\phi\, \phi(x_1) \ldots \phi(x_n) \exp(\mathrm{i}\mathcal{S})}{\int \mathscr{D}\phi \exp(\mathrm{i}\mathcal{S})}. \tag{2.3.1}$$

Here $\hat{\phi}$ represents an arbitrary (fermionic or bosonic) field, $\mathcal{T}$ is the time-ordering operator and $\mathscr{D}\phi$ stands for the path-integral measure, i. e., $\int \mathscr{D}\phi$ is a formal notation for an integral over all possible values of the fields at all space-time points. Moreover,

$$\mathcal{S} = \int \mathrm{d}^4 x\, \mathscr{L}(\phi, \partial_\mu \phi) \tag{2.3.2}$$

is the action of the given quantum field theory.

Introducing the generating functional,

$$\mathcal{Z}(J) = \int \mathscr{D}\phi \exp\left(\mathrm{i} \int \mathrm{d}^4 x [\mathscr{L} + \phi(x)J(x)]\right), \tag{2.3.3}$$

the n-point Green functions (2.3.1) are obtained by functional differentiation of $\mathcal{Z}$ with respect to the sources J:

$$\mathcal{G}^{(n)}(x_1, \ldots, x_n) = \frac{(-\mathrm{i})^n}{\mathcal{Z}(0)} \frac{\delta \mathcal{Z}(J)}{\delta J(x_1) \ldots \delta J(x_n)}\bigg|_{J=0}. \tag{2.3.1$'$}$$

Note, that due to the anticommutation relations fulfilled by fermions the corresponding classical fields appearing in the path integral must be represented by anticommuting numbers, the so-called Grassmann variables. Consequently, the sources for fermion fields are Grassmann-valued, too.

Gauge field theories involve integrations over (infinitely many) physically equivalent gauge field configurations. This requires introducing a gauge-fixing condition $G(A_\mu^a) = 0$. For $SU(N)$ Yang-Mills theories this is in general not sufficient to count each physical gauge configuration only once. There might arise so-called Gribov copies [11] for which there exists a gauge transformation U with $G(A^a) = G(A_U^a)$ (cf. Eq. (2.2.4)). In 1976 Faddeev and Popov (cf. [12]) showed how to incorporate properly the constraints dictated by gauge invariance into the path-integral formalism: multiplying the path integral Eq. (2.3.3) by 1 in the form

$$1 = \int \mathscr{D}\theta(x) \delta(G(A_U^a)) \det\left(\frac{\delta G(A_U^a)}{\delta \theta^b}\right),$$

where A_U^a denotes the transformed gluon field according to (2.2.4), and introducing Grassmann-valued fields χ and $\chi^\star$, known as *ghosts*, the determinant of the functional Jacobian $M^{ab}(x, y) := \frac{\delta G(A_U^a(x))}{\delta \theta^b(y)}$ can be written in the form

$$\det M = \mathrm{i} \int \mathscr{D}\chi \mathscr{D}\chi^\star \exp\left(-\mathrm{i} \int \mathrm{d}^4x \, \mathrm{d}^4y \, \chi^{a\star}(x) M^{ab}(x, y) \chi^b(y)\right).$$

Then the complete generating functional including all source terms (the sources $\xi, \xi^\star, \eta, \bar{\eta}$ for the ghosts and fermionic fields are Grassmann numbers while the sources $J^{a\mu}$ for the gluon fields are c-numbers) can be written for QCD as[9]

$$\mathcal{Z}(J, \xi, \xi^\star, \eta, \bar{\eta}) = \int \mathscr{D}A \, \mathscr{D}\chi \, \mathscr{D}\chi^\star \, \mathscr{D}\psi \, \mathscr{D}\bar{\psi}$$
$$\exp\left(\mathrm{i} \int \mathrm{d}^4x \left[\mathscr{L}_{\mathrm{QCD}} + A_\mu^a J^{a\mu} + \chi^{a\star}\xi^a + \xi^{a\star}\chi^a + \bar{\psi}\eta + \bar{\eta}\psi\right]\right), \tag{2.3.3$'$}$$

where the full QCD Lagrangian reads

$$\mathscr{L}_{\mathrm{QCD}} = \bar{\psi}(\mathrm{i}\gamma^\mu D_\mu - \hat{m})\psi - \frac{1}{2}\mathrm{Tr}(G_{\mu\nu}G^{\mu\nu}) - \frac{1}{2\lambda}(\partial^\mu A_\mu^a)^2 + (\partial^\mu\chi^{a\star})D_\mu^{ab}\chi^b. \tag{2.2.1$'$}$$

The third term on the right-hand side is the gauge-fixing term, the last term involves the ghost fields.

From the generating functional (2.3.3$'$) one can derive fermion and gluon propagators through functional differentiation. We have already seen that in QCD gluons do not only interact with

---

[9]This is, actually, the result for a generalized Lorentz gauge, $G(A_\mu^a) = \partial^\mu A_\mu^a(x) - \omega^a(x)$, with ω^a some functions chosen as a constraint. Only in such covariant gauges ghosts originate. If we had chosen an axial gauge, i. e., $G(A^a) = A_3^a = 0$, no ghost fields would have arisen.

quarks but also with each other. Therefore, we split the QCD Lagrangian into a free and an interaction part, i. e., we write $\mathscr{L}_{\text{QCD}} =: \mathscr{L}_0 + \mathscr{L}_I$. Then the generating functional can be decomposed as follows [9]:

$$\mathcal{Z}(J, \xi, \xi^\star, \eta, \bar{\eta}) = \exp\left[\mathrm{i} \int \mathrm{d}^4x\, \mathscr{L}_I\left(-\mathrm{i}\frac{\delta}{\delta J}, -\mathrm{i}\frac{\delta}{\delta\xi^{a\star}}, +\mathrm{i}\frac{\delta}{\delta\xi^a}, -\mathrm{i}\frac{\delta}{\delta\bar{\eta}}, +\mathrm{i}\frac{\delta}{\delta\eta}\right)\right] \mathcal{Z}_0(J, \xi, \xi^\star, \eta, \bar{\eta})\,.$$
(2.3.4)

Expanding the exponential in Eq. (2.3.4) finally leads to the perturbative expansion of QCD.

To conclude this section we point out that the resulting loop expansion is equivalent to an expansion in powers of the small quantity $\hbar$ around the classical theory. To see this one has to restore a factor $\frac{1}{\hbar}$ in the exponential of the path integral in (2.3.4). Then it is clear that each vertex contributes a factor $\hbar^{-1}$ and that each propagator contributes a factor $\hbar$ to a general graph in n-th order perturbation theory.

2.3.2 Renormalization Group Equations and Asymptotic Freedom

In order to deal with divergences that appear in loop corrections to Green functions the theory has to be regularized in order to isolate the singularities. Many different regularization schemes have been invented, each of them, however, exhibits some deficiencies: the (quite intuitive) cutoff method destroys translation and gauge invariance of the theory; the Pauli-Villars scheme is applicable to quantum electrodynamics (QED) but it ruins gauge invariance of non-Abelian theories; lattice calculations do not maintain translation and Lorentz invariance (cf. Sect. 2.4.3). Since gauge invariance is essential for QCD the most popular regularization scheme is dimensional regularization. It destroys only scale invariance (for problems which concern the extension of the matrix γ_5 to arbitrary dimensions we refer to the original paper by 't Hooft and Veltman [13]). In this regularization scheme diagrams are evaluated in $D = 4 - 2\varepsilon$ space-time dimensions and singularities can be extracted as poles in ε. From simple dimensional considerations it follows that the Lagrangian $\mathscr{L}$ has mass dimension D. Consequently, the dimension of the coupling constant g in $D = 4 - 2\varepsilon$ dimensions is equal to ε. Demanding that g should still be dimensionless in D dimensions one has to make the replacement

$$g \to g\mu^\varepsilon\,,$$

where μ is an arbitrary scale with the dimensions of mass. The appearance of the scale μ has profound impact on the theory.

In order to get rid of the divergences in Green functions the fields and parameters in the Lagrangian are renormalized according to

$$\begin{aligned} A_{0,\mu}^a &= Z_3^{1/2} A_\mu^a & \psi &= Z_\psi^{1/2}\psi \\ g_0 &= Z_g g\mu^\varepsilon & m_0 &= Z_m m\,. \end{aligned}$$
(2.3.5)

The index "0" denotes the unrenormalized ("bare") quantities, A_μ^a and ψ are renormalized fields,

g is the renormalized QCD coupling and m the renormalized quark mass. The (divergent) factors Z_i are called renormalization constants and have to be chosen such that all divergences disappear from the Green functions once expressed in terms of renormalized quantities. It is important to stress that the bare quantities do not depend on the scale μ. This implies that the renormalized quantities do depend on μ; in particular, the renormalization constants Z_i are μ dependent.

The explicit form of the renormalization constants depends on the renormalization scheme applied. A frequently used scheme is the modified minimal subtraction ($\overline{\text{MS}}$) scheme, in which not only divergent parts are subtracted but also finite terms that result throughout the calculations in D dimensions. In this scheme renormalization constants Z_i do not depend on masses. In this work all quantities which depend on the scale μ are understood to be calculated within the $\overline{\text{MS}}$ scheme.

The μ dependence of the renormalized coupling constant g (and of the quark mass) is governed by the renormalization group equations. These equations are derived easily from Eq. (2.3.5) noting that the bare quantities are μ independent:

$$\frac{\mathrm{d}g(\mu)}{\mathrm{d}\ln\mu} = -\varepsilon g + \beta(g)\,, \qquad (2.3.6)$$

where the (Gell-Mann–Low) β function is given by

$$\beta(g) = -\frac{g}{Z_g}\frac{\mathrm{d}Z_g}{\mathrm{d}\ln\mu}\,. \qquad (2.3.7)$$

This means that the μ dependence of the QCD coupling constant can be calculated from the vertex renormalization constant Z_g. From the leading-order contribution to the quark-gluon vertex one obtains

$$\beta(g) = -\beta_0\frac{g^3}{16\pi^2}$$

with $\beta_0 = (11N_c - 2N_f)/3$, where $N_c = 3$ and $N_f = 3$ is the number of quark flavors. Using this expression and setting $\alpha_s(\mu) := \frac{g^2(\mu)}{4\pi}$ allows one to integrate the renormalization group equation (2.3.6), with the leading-order result for the running QCD coupling:

$$\alpha_s(\mu) = \frac{4\pi}{\beta_0 \ln\left(\frac{\mu^2}{\Lambda_{\overline{\text{MS}}}^2}\right)}\,. \qquad (2.3.8)$$

As long as $\beta_0 > 0$ (for $N_c = 3$, $N_f \le 16$) the running coupling decreases logarithmically at large momentum scales (or short distances) so that the asymptotic freedom of QCD is realized. This is a remarkable result; the physical relevance to strong interactions was first discovered and understood by Gross and Wilczek [14, 15] and Politzer [16, 17]. Asymptotic freedom is a peculiar feature of renormalizable Yang-Mills theories, while not present in Abelian field theories like QED. In QED the coupling strength diminishes at larger distances owing to electrical charge-screening effects. In QCD one has the contrary, namely a net antiscreening effect of color charges.

2.4 Aspects of Nonperturbative QCD

After the discussion of the perturbative (high-momentum) regime of QCD we explore now the vacuum structure of QCD and nonperturbative effects, following loosely the lines of Ref. [18]. At energies and momenta below $Q < 1\,\mathrm{GeV}$, where the running coupling strength $\alpha_s(Q)$, Eq. (2.3.8), is of order one, a systematic expansion in powers of α_s is no longer possible: the region where perturbative QCD breaks down is called *nonperturbative QCD*. There exist several strategies how to approach this region; some of them (lattice QCD and Dyson-Schwinger equation) will be described and used in this work, for the other (e. g., QCD sum rules, chiral perturbation theory) we refer the reader to the literature.

2.4.1 Conservation Laws and PCAC

Symmetries of the QCD Lagrangian $\mathscr{L}_{\mathrm{QCD}}$, Eq. (2.2.1), have already been discussed in Sect. 2.2. Let now $N_f \geq 2$ be the number of flavors[10]. We mentioned, that the invariance of $\mathscr{L}_{\mathrm{QCD}}$ under the U(1)$_\mathrm{V}$ transformation implies a conserved baryon-number current, $\frac{1}{3}\bar{\psi}\gamma^\mu\psi$, with its conserved charge, the (properly normalized) baryon number

$$B = \frac{1}{3}\int \mathrm{d}^3x\,\psi^\dagger\psi\,. \tag{2.4.1}$$

Moreover, separate global chiral SU(N_f)$_\mathrm{R}$ × SU(N_f)$_\mathrm{L}$ transformations acting independently and the right- and left-handed quark fields, $\psi_{\mathrm{R,L}} = \frac{1}{2}(1 \pm \gamma_5)\psi$ respectively, leave $\mathscr{L}_{\mathrm{QCD}}$ with vanishing current quark masses invariant. This invariance under chiral symmetry implies $N_f^2 - 1$ conserved Noether vector currents,

$$J_{\mathrm{V},a}^\mu(x) = \bar{\psi}(x)\gamma^\mu\frac{\lambda_a}{2}\psi(x)\,, \qquad \text{with} \quad \partial_\mu J_{\mathrm{V},a}^\mu(x) = 0\,, \tag{2.4.2}$$

and $N_f^2 - 1$ conserved Noether axial currents,

$$J_{\mathrm{A},a}^\mu(x) = \bar{\psi}(x)\gamma^\mu\gamma_5\frac{\lambda_a}{2}\psi(x)\,, \qquad \text{with} \quad \partial_\mu J_{\mathrm{A},a}^\mu(x) = 0\,. \tag{2.4.3}$$

The λ_a stand for the generators of the SU(N_f) Lie algebra. The corresponding charges are

$$Q_\mathrm{V}^a = \int \mathrm{d}^3x\,\psi^\dagger(x)\frac{\lambda_a}{2}\psi(x)\,, \qquad Q_\mathrm{A}^a = \int \mathrm{d}^3x\,\psi^\dagger(x)\gamma_5\frac{\lambda_a}{2}\psi(x)\,. \tag{2.4.4}$$

If the chiral SU(N_f) × SU(N_f) symmetry is explicitly broken by small but finite current quark masses $\hat{m} = \mathrm{diag}(m_1, m_2, \ldots, m_{N_f})$, the divergence of the axial current becomes

$$\partial_\mu J_{\mathrm{A},a}^\mu(x) = \mathrm{i}\,\bar{\psi}\left\{\hat{m}, \frac{\lambda_a}{2}\right\}\gamma_5\psi\,. \tag{2.4.5}$$

This produces sources of pseudoscalar quark-antiquark pairs. This is the microscopic basis of

---

[10]Since this work deals both with $N_f = 2$ and $N_f = 3$ we leave the number of flavors at this stage unspecified.

the *partially conserved axial current* PCAC hypothesis that plays a key role in weak interactions of the nucleon and low-energy pion-nucleon dynamics. Important low-energy theorems involving pions (the Goldberger-Treiman and Gell-Mann–Oakes–Renner relations) are a consequence of the PCAC relation (2.4.5).

2.4.2 The QCD Vacuum (I):
Spontaneous Breaking of Chiral Symmetry

We return now to the case of vanishing current quark masses (the chiral limit) and consider the special case $N_f = 3$. There is plenty of evidence from hadron spectroscopy that the chiral $SU(3)_L \times SU(3)_R$ symmetry group is spontaneously broken to the flavor group $SU(3)_f \equiv SU(3)_V$ (for vanishing quark masses). For dynamical reasons of nonperturbative origin, the ground state $|0\rangle$ of QCD is symmetric only under the subgroup $SU(3)_V$ generated by the vector charges Q_V^a. If the ground state were symmetric under $SU(3)_L \times SU(3)_R$, both vector and axial charge operators would annihilate (2.4.4) the vacuum:

$$Q_V^a|0\rangle = Q_A^a|0\rangle = 0 \qquad \text{(Wigner, Weyl)}. \qquad (2.4.6)$$

This is the so-called Wigner-Weyl realization of chiral symmetry with a trivial vacuum. It would imply that parity doublets appear in the hadron spectrum. Consequently, the spectra of, e. g., pseudoscalar ($J^P = 0^-$) and scalar ($J^P = 0^+$) mesonic excitations should be identical. This degeneracy is, however, not observed in nature: the pion mass ($m_\pi \approx 140\,\text{MeV}$) is well separated by a gap from that of the a_0 meson ($m_{a_0} \approx 980\,\text{MeV}$). Hence, in order to be compatible with experimental results, only $Q_V^a|0\rangle = 0$ can be maintained, and we conclude $Q_A^a|0\rangle \neq 0$. This is the so-called Nambu-Goldstone realization of chiral symmetry. In this case, according to the Goldstone theorem there exists a massless Goldstone boson

$$|\phi^a\rangle = Q_A^a|0\rangle \qquad \text{(Nambu, Goldstone)} \qquad (2.4.7)$$

which is energetically degenerate with the ground state $|0\rangle$ and which carries the quantum numbers of the corresponding axial charges. Hence for $N_f = 3$, the Nambu-Goldstone bosons are the eight pseudoscalar mesons (pions, kaons and eta).

Since phase transitions are best characterized by the behavior of an order parameter, we now introduce the order parameter relevant for the chiral phase transition: the chiral condensate. Spontaneous chiral symmetry breaking is accompanied by a qualitative rearrangement of the QCD ground state: the vacuum is populated by a condensate of scalar quark-antiquark pairs, characterized by a nonzero expectation value of the composite operator $\bar{\psi}\psi$; for $N_f = 3$ we have:

$$\langle \bar{\psi}\psi \rangle := \langle 0|\bar{\psi}\psi|0\rangle = \langle 0|\bar{u}u|0\rangle + \langle 0|\bar{d}d|0\rangle + \langle 0|\bar{s}s|0\rangle. \qquad (2.4.8)$$

This quantity is called the chiral condensate and its exact definition is

$$\langle \bar{\psi}\psi \rangle = -\mathrm{i}\,\mathrm{Tr}\,\lim_{y \to x^+}\left[S_\mathrm{F}(x,y) - S_\mathrm{F}^{(0)}(x,y)\right]. \qquad (2.4.9)$$

Herein, $S_\mathrm{F}(x,y) = -\mathrm{i}\langle 0|\mathcal{T}\psi(x)\bar{\psi}(y)|0\rangle$ is the full quark propagator ($\mathcal{T}$ denotes the time-ordering operator) and $S_\mathrm{F}^{(0)}(x,y)$ is the perturbative quark propagator. From the subtraction in Eq. (2.4.9) it is manifest that $\langle \bar{\psi}\psi \rangle$ is a purely nonperturbative quantity.[11] Writing $\bar{\psi}\psi = \bar{\psi}_\mathrm{R}\psi_\mathrm{L} + \bar{\psi}_\mathrm{L}\psi_\mathrm{R}$, one sees that $\bar{\psi}\psi$ is not chirally invariant. Hence, the appearance of a nonvanishing chiral condensate signals that the ground state is restructured such that it has lost the chiral symmetry of the underlying Lagrangian.

Low-Energy Theorems

Since we have now investigated conservation laws following from the QCD Lagrangian and the vacuum structure of QCD we are now able to derive some fundamental theorems that have to hold in the model that we use in this work. From hadron spectroscopy we deduced that chiral symmetry is broken down to $\mathrm{SU}(3)_\mathrm{V}$ giving rise to pseudoscalar Goldstone fields $|\pi^a\rangle$, compare Eq. (2.4.7). The Goldstone theorem then states

$$\langle 0|J_{\mathrm{A},a}^\mu(x)|\pi_b(p)\rangle = \mathrm{i}p^\mu f_0 \delta_{ab}\mathrm{e}^{-\mathrm{i}p\cdot x}. \qquad (2.4.10)$$

f_0 is the *pion decay constant* in the chiral limit. The physical value of the pion decay constant is deduced from charged pion decay (cf. [19]) and given by

$$f_\pi = (92.4 \pm 0.3)\,\mathrm{MeV}. \qquad (2.4.11)$$

The difference between f_0 and f_π is a correction linear in the quark mass m_0.

Nonvanishing quark masses shift the mass of the Goldstone boson from zero to the observed value of the physical pion mass m_π. Using the PCAC relation Eq. (2.4.5), one obtains in the $N_\mathrm{f} = 2$ case[12] for a selected isospin component with $a = 1$

$$\partial_\mu J_{\mathrm{A},1}^\mu = (m_u + m_d)\bar{\psi}\mathrm{i}\gamma_5\frac{\tau_1}{2}\psi.$$

In addition, applying the canonical anticommutation relations for fermionic fields and $[\hat{A},\hat{B}] = \{\hat{A},\hat{B}\} - 2\hat{B}\hat{A}$ one shows for $\hat{P}_a := \bar{\psi}(x)\gamma_5\tau_a\psi(x)$ the relation

$$\left[\hat{Q}_\mathrm{A}^a, \hat{P}_b\right] = -\delta_{ab}\bar{\psi}\psi.$$

Then one can combine these two relations to obtain

$$\left\langle 0\left|\left[\hat{Q}_\mathrm{A}^1, \partial_\mu J_{\mathrm{A},1}^\mu\right]\right|0\right\rangle = -\frac{\mathrm{i}}{2}(m_u + m_d)\langle \bar{u}u + \bar{d}d\rangle.$$

---

[11] The condensates $\langle \bar{u}u\rangle$, $\langle \bar{d}d\rangle$ etc. depend on the renormalization scale; therefore, they are not observables in QCD. Only the products $m_\mathrm{f}\langle \bar{q}_\mathrm{f}q_\mathrm{f}\rangle$ are invariants of the renormalization group.

[12] In the two-flavor case the generators of $\mathrm{SU}(N_\mathrm{f})$ are given by the isospin Pauli matrices $\{\frac{\tau_1}{2}, \frac{\tau_2}{2}, \frac{\tau_3}{2}\}$.

The expression on the left-hand side can easily be calculated from Eq. (2.4.10) after inserting a complete set of pseudoscalar states in the commutator. Truncating this set by the one-pion states $|\pi_a\rangle$ we obtain

$$m_\pi^2 = -\frac{1}{f_0^2} m \langle \bar{\psi}\psi \rangle + \mathcal{O}(m^2), \qquad (2.4.12)$$

where we have assumed the isospin limit, i.e., $m_u \approx m_d := m$. This result is the celebrated Gell-Mann–Oakes–Renner relation (cf. [20]) that is based entirely on the symmetries and symmetry breaking patterns of QCD.

Another important relation can be obtained when evaluating the matrix element of the weak axial-vector current $J_{A,1}^\mu + iJ_{A,2}^\mu$ between nucleon states $|p\rangle, |n\rangle$. This matrix element is measured in neutron β decay. The PCAC relation connects this to the charged-pion matrix element $\langle p|\pi^+|n\rangle$ and one recovers the Goldberger-Treiman relation [21]

$$f_\pi g_{\pi NN} = M_N g_A, \qquad (2.4.13)$$

where $g_{\pi NN} = 13.2 \pm 0.1$ is the pion-nucleon-nucleon coupling constant, $M_N = 0.939\,\text{GeV}$ is the nucleon mass and $g_A = 1.267 \pm 0.004$ is the nucleon-axial-vector coupling constant. The Goldberger-Treiman relation is thus better than $5\,\%$.

2.4.3 Lattice QCD

Asymptotic freedom allows for a perturbative treatment of QCD at high energies and momenta. Until 1974 all predictions of QCD were restricted to this regime. In order to study the low-energy, or long-distance, properties of QCD Kenneth Wilson [22] introduced in 1974 the *lattice gauge theory*. In lattice gauge theories one discretizes the space-time continuum, which provides a natural cutoff scheme (lattice regularization): wavelengths shorter than twice the lattice spacing a have no meaning; equivalently one can state that momenta k are restricted to the first Brillouin zone, i.e., $|k| \leq \pi/a$. One usually introduces the lattice in field theory by performing a Wick rotation to Euclidean space, i.e., $x_0 \to -ix_4$, $x_i \to x_i, i \in \{1,2,3\}$, and by quantizing the theory via the path-integral formalism (Euclidean lattice formulation[13]). One of the advantages of lattice quantum field theory is that the path integrals can be given a precise meaning since one is dealing with only a finite number of degrees of freedom. A nontrivial task is, however, the removal of the lattice structure and its artifacts by studying the continuum limit.

Considerable freedom exists in the lattice formulation: one is free to add to the Lagrangian terms that do not contribute in the continuum limit. Wilson took advantage of this freedom when writing down the lattice formulation for gauge theories which keep local gauge invariance as an exact symmetry. In the QCD Lagrangian $\mathcal{L}_{\text{QCD}}$ the partial derivative is replaced by the covariant one incorporating the gauge field. This is, actually, one particular (the infinitesimal) manifestation of the more general concept of *parallel transport*[14]: suppose we want to gauge

---

[13]There exists another method, the Hamiltonian lattice formulation of Kogut and Susskind [23], in which only the spatial dimensions are discretized in a Minkowski space-time.

[14]The reader could be familiar with this argument from differential geometry, or, more specifically, from its application in general relativity.

(with gauge group $\mathcal{G}$) a nonlocal operator $\hat{\mathcal{O}}(x, y) = \hat{\mathcal{A}}(x)\hat{\mathcal{B}}(y)$; it turns out that the correct prescription is given by

$$\hat{\mathcal{A}}(x)\hat{\mathcal{B}}(y) \underset{\mathcal{G}}{\rightarrow} \hat{\mathcal{A}}(x)U(x, y)\hat{\mathcal{B}}(y) , \tag{2.4.14}$$

with the parallel-transport operator

$$U_{\mathscr{C}}(x, y) = \mathcal{P}\left\{ \exp\left[ig \int_{\mathscr{C}} \frac{\lambda_a}{2} A_\mu^a(s) \, \mathrm{d}s^\mu \right] \right\} , \tag{2.4.15}$$

where $\mathscr{C}$ denotes an arbitrary path connecting y and x, $\mathcal{P}$ is the path-ordering operator, the λ_a are the generators of the gauge group $\mathcal{G}$ and the A_μ^a describe the corresponding gauge fields. Since all lattice sites are separated at least by the (finite) lattice spacing a, the gauge action must be constructed out of (discretized) parallel-transport operators $U(n, m)$, where n, m now denote lattice sites and U is called *link variable*. It turns out that the lattice action for the gluon field is given by a sum over *plaquettes* U_p,

$$U_p = U(n, n + \hat{\mu})U(n + \hat{\mu}, n + \hat{\mu} + \hat{\nu})U(n + \hat{\mu} + \hat{\nu}, n + \hat{\nu})U(n + \hat{\nu}, n) ,$$

where $\hat{\mu}, \hat{\nu}$ denote unit vectors on the lattice. The gauge part of an SU(N) lattice action is then given by [24]

$$S_{\text{gauge}}^{(\text{SU}(N))} = \frac{2N}{g_0^2} \sum_p \left[1 - \frac{1}{2N} \mathrm{Tr}\left(U_p + U_p^\dagger \right) \right] \tag{2.4.16}$$

where the sum extends over all distinct (path-ordered) plaquettes on the lattice, and g_0 denotes the bare coupling strength.

We have stressed here the concept of parallel transport, because we will use in the second part of this work that $U_{\mathscr{C}}(x, x)$, the Wilson loop, serves as an order parameter for confinement (see Sect. 5.2).

So far, it was straightforward to describe a (pure) gauge theory on the lattice. Problems arise when describing fermions on the lattice: when naively discretizing the (Euclidean) Dirac action,

$$S_{\text{Dirac}} = \int \mathrm{d}^4 x \bar{\psi}(x)(\gamma_\mu D^\mu + \hat{m})\psi(x) ,$$

by replacing D^μ with a covariant difference operator, one obtains an action that describes 16 degenerate types of fermions ("tastes"). This is the famous fermion doubling problem which is strongly related to the axial anomaly on the lattice. Since one wants to describe only one fermion, one has to get rid of the other 15 tastes. There are several possibilities. The first (and oldest) is due to Wilson, who added some term (that vanishes in the continuum limit) to the action to change the propagator so that it has only one pole (only one of the 16 fermions remains massless, the other 15 gain masses that are inversely proportional to the lattice spacing). The Wilson solution to the doubling problem is easy to understand, implement and compute with. But it suffers from a number of flaws. The first is that the Wilson term breaks chiral symmetry. Another (technical) problem with Wilson quarks is that the Wilson term introduces a linear error in the lattice spacing a that can be very large. That is the reason why one has been looking for other approaches: in the "staggering" method the four spin components of the Dirac spinor are put on different sites of the lattice. This procedure reduces the number of tastes from 16 to four[15]. The major advantage of the staggered-fermion over the Wilson-fermion approach is that the action of the first one preserves a continuous U(1) × U(1) symmetry (in the chiral limit) which is a remnant of the chiral symmetry group. Although neither one of the approaches incorporates chiral symmetry properly (we will see the reason for this shortly), one can nevertheless use the staggered-fermion formulation to study the spontaneous breakdown of this remaining lattice symmetry and the associated Nambu-Goldstone phenomenon. One of the drawbacks of this method is that since the different components of the staggered quark field live on different lattice sites, they experience a slightly different gauge field, which leads to a breaking of their naive degeneracy and, hence, to Nambu-Goldstone bosons with nonvanishing mass even in the chiral limit.[16]

The fact that chiral symmetry is not preserved by lattice actions is a possible consequence of the Nielsen-Ninomiya no-go theorem [26]. It states that it is impossible to have a chirally invariant, doubler-free, local, translation invariant, real bilinear fermion action on the lattice. Since locality, translation invariance and hermiticity should always be fulfilled by the action, on first sight, one has to abandon either chiral symmetry or freedom from doublers. It is, however, possible to circumvent the Nielsen-Ninomiya theorem: one possibility is to realize the four-dimensional theory of a chiral fermion by dimensional reduction from a five-dimensional theory (domain-wall or overlap fermions). With a finite extent L_s in the fifth direction, there will necessarily be another domain wall with opposite orientation, on which a massless chiral fermion of opposite chirality will live, thus fulfilling both the Nielsen-Ninomiya theorem in the five-dimensional theory and ensuring the mutual cancelation of the chiral anomalies stemming from either fermion. To be precise, chiral symmetry in the domain-wall formalism is preserved only for large L_s. A finite separation L_s between the walls with opposite chirality gives rise to a residual chiral symmetry breaking that can be strongly suppressed, though, by taking L_s to be large. To leading order in an expansion in lattice spacing, the residual chiral symmetry breaking can be characterized by a single parameter, the residual mass m_{res}. This acts as an additive shift to the bare input quark mass m_0. Hence, the "physical" quark mass is given by $m_{res} + m_0$. The full continuum SU(N_f) × SU(N_f) chiral symmetry can be reproduced by choosing L_s sufficiently large, even at finite lattice spacing. This comes, however, with an approximate factor of L_s increase in computational cost.

Lattice calculations are performed on large-scale computers using sophisticated numerical algorithms (Monte Carlo methods). Particularly time-consuming is the evaluation of the fermion determinant. Until the mid-nineties it was essentially impossible to simulate the full action on the lattice. This led people to drop the fermion determinant, using the so-called quenched approximation, that amounts to neglecting dynamical fermions in the simulation.

We will come back to lattice simulations in the second part of this work (Sect. 4.3), when dealing with the finite temperature description of QCD.

---

[15]The number of fermions can be reduced to one when artificially taking the fourth root of the fermion determinant. This rooting procedure is, however, still under lively debate (cf. [25]).

[16]Improved staggered lattice actions are known that do not suffer from such pathologies.

2.5 Dyson-Schwinger Formalism

Apart from lattice simulations, the Dyson-Schwinger formalism provides an alternative approach to QCD. The Dyson-Schwinger equations (DSE) are a nonperturbative tool for analyzing a quantum field theory starting from the theory's generating functional in Euclidean space-time. From this one obtains an enumerably infinite set of coupled integral equations, the so-called DSE tower. Since it is not possible to solve all of them, one has to truncate this tower at some order. In theories with elementary fermions, the simplest of the DSEs is the gap equation, which already incorporates the dynamical (chiral) symmetry breaking. From this one notes that a self-consistent solution of the DSEs comprises nonperturbative effects that would not be included in a systematic perturbative expansion.

We derive the fermion gap equation for QCD and determine its solution, which is actually a two-point function. This procedure will guide us in developing the model that is the major subject of this work and it will serve as a consistency check whenever simplifications are made.

2.5.1 Euclidean Action

From now onwards we work in Euclidean space-time without exception. Euclidean space has already been encountered when describing lattice QCD in the previous section. The following conventions will be used: for four-vectors a, b we have

$$a \cdot b := a_\mu b_\nu \delta_{\mu\nu} := \sum_{i=1}^{4} a_i b_i \,. \tag{2.5.1}$$

A distinction between co-variant and contra-variant indices is not necessary anymore. With this definition, spacelike vectors x_μ have $x^2 > 0$. Furthermore, a four-vector x_μ can be Wick-rotated from Minkowski to Euclidean space according to

$$(x_0, \vec{x}) \rightarrow (\vec{x}, x_4) \,, \qquad \text{with } x_4 = \mathrm{i}x_0 \,. \tag{2.5.2}$$

Finally, we need to define the Dirac matrices in Euclidean space. Here we use the following convention:

$$\gamma_4 = \mathrm{i}\gamma_0 \tag{2.5.3}$$

(and all other Dirac matrices unchanged: $\gamma_{\mathrm{E},i} = \gamma_{\mathrm{E}}^i := \gamma^i$ for $i \in \{1, 2, 3\}$). Then, the matrices fulfill the algebra

$$\{\gamma_\mu, \gamma_\nu\} = -2\delta_{\mu\nu} \,. \tag{2.5.4}$$

In particular, one obtains

$$\slashed{\partial} = \gamma_4 \partial_4 + \vec{\gamma} \cdot \nabla \,. \tag{2.5.5}$$

Using these conventions the Euclidean QCD action can be calculated from the Minkowskian expression, $\mathcal{S}_{\mathrm{M}}^{\mathrm{QCD}} = \int \mathrm{d}^4 x_{\mathrm{M}} \mathcal{L}_{\mathrm{QCD}}$ (cf. Eq. (2.2.1)), replacing the expressions properly in Euclidean space:

$$S_{\mathrm{E}}^{\mathrm{QCD}} = \int \mathrm{d}^4 x \left[\frac{1}{2} \operatorname{Tr} \left(G_{\mu\nu} G_{\mu\nu} \right) + \frac{1}{2\lambda} \left(\partial_\mu A_\mu^a \right)^2 - (\partial^\mu \chi^{a\star}) D_\mu^{ab} \chi^b + \sum_{f \in \mathrm{flavors}} \bar{\psi}_f \left(-\mathrm{i}\partial\!\!\!/ + m_f + g A\!\!\!/ \right) \psi_f \right].$$
$$(2.5.6)$$

From this the generating functional follows as in Eq. (2.3.3′):

$$\mathcal{Z}(J, \xi, \xi^\star, \eta, \bar{\eta}) = \int \mathscr{D}A \, \mathscr{D}\chi \, \mathscr{D}\chi^\star \, \mathscr{D}\psi \, \mathscr{D}\bar{\psi}$$
$$\exp\left(-S_{\mathrm{E}}^{\mathrm{QCD}} \right) \exp\left(\mathrm{i} \int \mathrm{d}^4 x \left[A_\mu^a J^{a\mu} + \chi^{a\star} \xi^a + \xi^{a\star} \chi^a + \bar{\psi}\eta + \bar{\eta}\psi \right] \right).$$
$$(2.5.7)$$

The Dyson-Schwinger equations are derived from the generating functional demanding that its derivative with respect to the fields is zero. We will see this explicitly in the next subsection.

2.5.2 Dyson-Schwinger Equation for the Quark Propagator

Consider the quark Dyson-Schwinger equation, or *gap equation*, which is useful when studying dynamical chiral symmetry breaking. The quark DSE in momentum space is given by

$$(2.5.8)$$

$$S(p)^{-1} = -p\!\!\!/ + m_0 - g^2 \int \frac{\mathrm{d}^4 q}{(2\pi)^4} D_{\mu\nu}(p - q) \frac{\lambda_a}{2} \gamma_\mu S(q) \frac{\lambda_a}{2} \Gamma_\nu(q, p)$$

where $S(p)$ is the *dressed* quark propagator, $D_{\mu\nu}(k)$ is the dressed gluon propagator[17] (curly line with the filled circle), $\Gamma_\nu(q, p)$ is the dressed quark vertex (full circle), $\frac{\lambda_a}{2}\gamma_\mu$ is the one-particle irreducible vertex (open circle), m_0 is the (bare) current quark mass and $-p\!\!\!/ + m_0$ is the inverse free fermion propagator (solid line). Furthermore, we use the propagators expressed in *Landau gauge*, $\lambda = 0$.[18]

The solution of Eq. (2.5.8) has the form

$$S(p)^{-1} = -p\!\!\!/ + M(p^2),$$
$$(2.5.9)$$

where $M(p^2)$ is the (dynamically generated) quark mass function. In the chiral limit, defined as $m_u = m_d = m_s = 0$, there is no perturbative contribution to the mass function $M(p^2)$ in Eq. (2.5.9). This means that $M(p^2) \neq 0$ in the chiral limit has a nonperturbative origin from the QCD vacuum and is hence related to spontaneous chiral symmetry breaking. In fact, inserting Eq. (2.5.9) in definition Eq. (2.4.9), one sees that $M(p^2) \neq 0$ in the chiral limit is only possible if the quark condensate is nonzero, which serves, as described in Sect. 2.4.2, as an order parameter

---

[17]The dressed gluon propagator is diagonal in color space because we do not allow for anisotropies in color space.

[18]Landau gauge is convenient because it is a fixed point of the renormalization group, i. e., $\lambda = 0$ at all orders in perturbation theory.

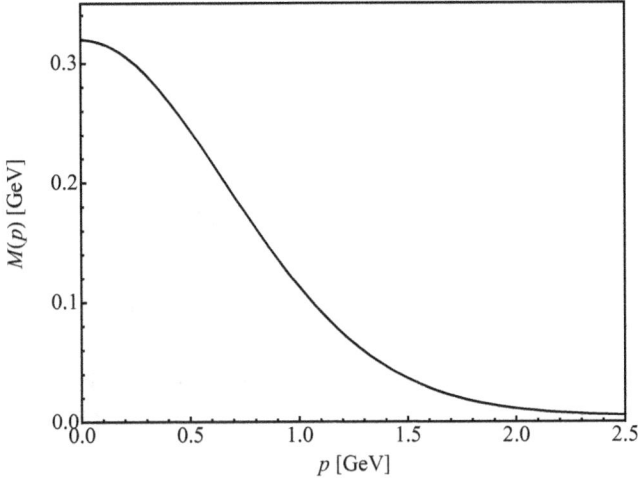

FIGURE 2.1: Solution to the gap equation Eq. (2.5.12) using a Gaussian ansatz for the dressed gluon propagator. The nonvanishing dynamical quark mass function $M(p)$ indicates that chiral symmetry is spontaneously broken.

for spontaneous chiral symmetry breaking.

Consider now a schematic solution of the gap equation Eq. (2.5.8) in the two-flavor case. Since the quark DSE is an integral equation, all of its elements must be known at all values of their momentum arguments, i.e., not just in the perturbative domain, but also in the infrared. The gluon propagator and the quark-gluon vertex fulfill their own DSE, which couples the quark DSE to the other members of the tower of equations (diagrammatic representation in Eq. (2.5.8)). Therefore we employ an ansatz for the dressed gluon propagator $D_{\mu\nu}(k)$ and the vertex function $\Gamma_\nu(q,p)$. First assume that the gluon propagator is diagonal in Dirac space, i.e., we set

$$g^2 D_{\mu\nu}(k) = \frac{3}{2} N_c G \, \mathcal{G}(k) \, \delta_{\mu\nu} \, ; \tag{2.5.10}$$

this is the so-called "Abelian approximation". We choose $\mathcal{G}$ such that it is normalized to $\mathcal{G}(k = 0) = 1$. We have introduced a constant G with mass-dimension -2; the prefactor has been chosen for practical reasons when comparing the results obtained in this section with those of Chapt. 3. Furthermore, we use the "rainbow truncation" for the dressed vertex function:

$$\Gamma_\nu(q,p) = \gamma_\nu \, . \tag{2.5.11}$$

Inserting this in the gap equation (2.5.8), using the general expression (2.5.9) and carrying out the algebra, we finally obtain

$$M(p) = m_0 + 8 G N_c \int \frac{\mathrm{d}^4 q}{(2\pi)^4} \, \mathcal{G}(p-q) \frac{M(q)}{q^2 + M^2(q)} \, . \tag{2.5.12}$$

There exist sophisticated parametrizations for the gluon function $\mathcal{G}$ in the literature (cf., e. g., Ref. [27]). For schematic simplicity, choose $\mathcal{G}(k) = \exp(-k^2/\beta^2)$, a Gaussian normalized such that $\mathcal{G}(0) = 1$. We will see in Chapt. 3 that this ansatz actually turns out to be quite reasonable. Eq. (2.5.12) then depends on two constants, namely the coupling constant, G, and the width of the Gaussian, β. These parameters, in addition to the current quark mass $m_0 = m_u = m_d$ which might vary in the range $m_0 \in [3, 10]$ MeV, are fixed by reproducing the pion mass $m_\pi \approx 140$ MeV and the pion decay constant $f_\pi \approx 92$ MeV; formulas are taken from Ref. [28]. The solution of the gap equation (2.5.12) has been performed iteratively relying on Banach's fixed-point theorem. The resulting mass function is shown in Fig. 2.1. It is clear that with $M(0) \approx 300$ MeV, chiral symmetry is spontaneously broken. From the Gell-Mann–Oakes–Renner relation (2.4.12) we immediately get the chiral condensate $(m_0 = 3.7$ MeV$)$ $\langle \bar{u}u \rangle = \langle \bar{d}d \rangle \approx -(280$ MeV$)^3$.

We have shown here how the simplest Dyson-Schwinger equation can be solved using assumptions about the fully dressed gluon propagator and vertex function. The next logical step would be an extension of this formalism to three flavors and to finite temperature and densities. This is, however, not trivial and, in particular, the implementation of thermodynamics in the Dyson-Schwinger formalism is involved. The gap equation (2.5.12) as a simplified version of the DSE (2.5.8) is nonetheless useful for guidance in the modeling of QCD thermodynamics, the central topic of this work to which we now turn.

3 Nonlocal Nambu–Jona-Lasinio Models

In this chapter we establish contacts between the Dyson-Schwinger formalism, described and worked out in the previous chapter, Sect. 2.5, and the time-honored Nambu–Jona-Lasinio model. At the end we will have developed a model with a gap equation only slightly different from the DSE gap equation (2.5.8), which is, however, perfectly suited for a practicable implementation of thermodynamics and a direct comparison to studies within a Polyakov-loop-extended Nambu–Jona-Lasinio model performed over the last decade (see Chapt. 5). We consider both the two- and three-flavor cases.

3.1 The Local Nambu–Jona-Lasinio Model

The ground state of QCD spontaneously breaks chiral symmetry. Furthermore, the axial U(1) symmetry is anomalously broken by instanton effects (cf. Sect. 3.3.1). All these features are embodied in $\mathscr{L}_{\mathrm{QCD}}$, Eq. (2.2.1). In this chapter the aim is to construct a model Lagrangian that is simpler to work with than full QCD but nevertheless encodes all relevant aspects of the symmetry breaking pattern of QCD. A model fulfilling this requirement was proposed in 1961 by Y. Nambu and G. Jona-Lasinio (NJL) [29,30] already in the pre-QCD era. Their aim was, actually, to describe a mechanism that is able to give rise to the nucleon mass. Treating primary fermionic degrees of freedom like in the Bardeen-Cooper-Schrieffer theory (BCS theory) of superconductivity, Nambu and Jona-Lasinio obtained a gap equation determining the nucleon mass. Considering a simplified model of a nonlinear four-fermion interaction that preserves $\mathrm{SU}(2)_{\mathrm{L}} \times \mathrm{SU}(2)_{\mathrm{R}}$ symmetry, Nambu and Jona-Lasinio got, in addition, pseudoscalar zero-mass bound states of nucleon-antinucleon pairs which were identified with pions.

Once QCD was established as the theory of strong interactions, the NJL model has been interpreted as an effective theory of quarks. For $N_{\mathrm{f}} = 2$ quark flavors the minimal NJL Lagrangian reads[1]:

$$\mathscr{L}_{\mathrm{NJL}}^{2\,\mathrm{f}} = \bar{\psi}\left(-\mathrm{i}\partial\!\!\!/ + \hat{m}\right)\psi - \frac{G}{2}\left[\left(\bar{\psi}\psi\right)^2 + \left(\bar{\psi}\mathrm{i}\gamma_5\vec{\tau}\psi\right)^2\right] \tag{3.1.1}$$

where $\hat{m} = \mathrm{diag}(m_u, m_d)$ is the current quark mass matrix, ψ stands for the two-flavor quark fields, τ_k with $k \in \{1,2,3\}$ denote the isospin Pauli matrices and G is the NJL four-point coupling strength with mass dimension -2. The Lagrangian $\mathscr{L}_{\mathrm{NJL}}^{2\,\mathrm{f}}$ is invariant under $\mathrm{SU}(2)_{\mathrm{L}} \times \mathrm{SU}(2)_{\mathrm{R}} \times \mathrm{U}(1)_{\mathrm{V}}$ symmetry. The U(1) axial anomaly is explicitly broken already at the level of the Lagrangian. A corresponding Lagrangian for the $N_{\mathrm{f}} = 3$ flavor case (cf. Ref. [31]) has the

---

[1]Remember that we use exclusively Euclidean space notation.

form:

$$\mathscr{L}_{\mathrm{NJL}}^{3\,\mathrm{f}} = \bar{\psi}\left(-\mathrm{i}\partial\!\!\!/ + \hat{m}\right)\psi - \frac{G}{2}\sum_{a=0}^{8}\left[\left(\bar{\psi}\lambda_a\psi\right)^2 + \left(\bar{\psi}\mathrm{i}\gamma_5\lambda_a\psi\right)^2\right]$$
$$- H\left\{\det\left[\bar{\psi}(1+\gamma_5)\psi\right] + \det\left[\bar{\psi}(1-\gamma_5)\psi\right]\right\} \tag{3.1.2}$$

where λ_a with $a \in \{1,\ldots,8\}$ are the SU(3) flavor Gell-Mann matrices and $\lambda^0 = \sqrt{\frac{2}{3}}\mathrm{diag}(1,1,1)$; $\hat{m} = \mathrm{diag}(m_u, m_d, m_s)$ and ψ is the three-flavor quark field. det here denotes the determinant in flavor space, and G and H are coupling constants. The four-fermion symmetric term in Eq. (3.1.2) is invariant under U(3)$_\mathrm{L}$ × U(3)$_\mathrm{R}$, while the determinant term [32, 33], a genuine three-fermion interaction, is required to remove the U(1)$_\mathrm{A}$ symmetry that would otherwise be present. We have mentioned already in Sect. 2.2 that the anomalous breaking is the reason for the mass of the η' meson. Note that the two-flavor NJL Lagrangian (3.1.1) can be cast into the same form as the three-flavor Lagrangian (3.1.2) after a re-distribution of the terms:[2]

$$\mathscr{L}_{\mathrm{NJL}}^{2\,\mathrm{f}} = \bar{\psi}\left(-\mathrm{i}\partial\!\!\!/ + \hat{m}\right)\psi - \frac{G}{4}\sum_{i=0}^{3}\left[\left(\bar{\psi}\tau_i\psi\right)^2 + \left(\bar{\psi}\mathrm{i}\gamma_5\tau_i\psi\right)^2\right]$$
$$- \frac{G}{2}\left\{\det\left[\bar{\psi}(1+\gamma_5)\psi\right] + \det\left[\bar{\psi}(1-\gamma_5)\psi\right]\right\}. \tag{3.1.1$'$}$$

The new feature for three flavors is the additional independent coupling H.

At this stage, both the two- and three-flavor NJL Lagrangians are not confining. They should be applied only to properties for which confinement is not essential. The point interaction renders the model nonrenormalizable. A cutoff Λ_{NJL} is necessary to regulate the theory. This cutoff introduces a characteristic length scale for the interaction. Using Λ_{NJL} as a three-momentum cutoff, one typically gets $\Lambda_{\mathrm{NJL}} \approx 650\,\mathrm{MeV}$ (see, e. g., Refs. [34–38]).

3.2 Nonlocal Nambu–Jona-Lasinio Model for Two Flavors

The original NJL model effectively introduces a schematic interaction that is strong at quark momenta $|\vec{p}| \leq \Lambda_{\mathrm{NJL}}$ and turned off at $|\vec{p}| > \Lambda_{\mathrm{NJL}}$. Closer contact with QCD and its running coupling strength can be established in a generalized nonlocal NJL model [39, 40] to which we now turn. In this section we describe the two-flavor version of this nonlocal NJL model. The generalization to three flavors needs some more preparation and will be discussed in Sect. 3.3.

3.2.1 The Nonlocal Action

We start with the construction of the basic two-flavor action [39], $\mathcal{S}_{\mathrm{E}}^{2\,\mathrm{f}}$, that will be used in this work:

$$\mathcal{S}_{\mathrm{E}} = \mathcal{S}_0 + \mathcal{S}_{\mathrm{int}}^{(4)}, \tag{3.2.1}$$

---

[2]Here we define $\tau_0 := 1$, such that the relation $\mathrm{tr}\{\tau_i \cdot \tau_j\} = 2\delta_{ij}$ holds for all $i, j \in \{0, 1, 2, 3\}$.

where we separate the free part of the action,

$$S_0 = \int \mathrm{d}^4 x \, \bar{\psi} \left(-\mathrm{i}\partial\!\!\!/ + \hat{m} \right) \psi \,, \tag{3.2.2}$$

with the quark fields $\psi(x) = (\psi_u(x), \psi_d(x))^\top =: (u(x), d(x))^\top$ (following the notation in Eq. (2.2.1)) and $\hat{m} = \mathrm{diag}(m_u, m_d)$. In this work, we restrict ourself to the isospin symmetric case $m_u = m_d$.

Consider now a QCD effective interaction between two color currents $J_a^\mu(x) = \bar{\psi}(x)\gamma^\mu \frac{\lambda_a}{2}\psi(x)$ ($\lambda_a/2$ denote the generators of the SU(3) color gauge group). This interaction contributes a term of the following generic form to the action:

$$S_{\mathrm{int}} = -\int \mathrm{d}^4 x \int \mathrm{d}^4 y \, J_a^\mu(x) \, \mathcal{D}_{\mu\nu}(x-y) \, J_a^\nu(y) \,. \tag{3.2.3}$$

Here $\mathcal{D}_{\mu\nu}(x-y)$ is proportional to the gluonic field correlator (we recall that $\mathcal{D}_{\mu\nu}$ is diagonal in color space, excluding color anisotropies), compare Eq. (2.5.6). As discussed in Sects. 2.5 and 2.4.3, approximate solutions for this correlator can be obtained from Dyson-Schwinger calculations and from lattice QCD. We do not need to specify $\mathcal{D}_{\mu\nu}$ in all detail at this stage. Eq. (3.2.3) defines a nonlocal effective interaction between quarks. The range of the nonlocality is determined by the correlation length characteristic of the color exchange through gluon fields.

In order for our model to be compatible with the Dyson-Schwinger gap equation (2.5.8) we use the "Abelian approximation" (cf. Eq. (2.5.10)) for the leading behavior of $\mathcal{D}_{\mu\nu}$ here as well, i.e., $\mathcal{D}_{\mu\nu}(x-y) \sim \delta_{\mu\nu} \frac{G}{2} \mathcal{G}(x-y)$, with a distribution function $\mathcal{G}(x-y)$ of mass dimension 4 and an NJL-type four-fermion coupling strength G with mass dimension -2. A Fierz exchange transformation then leads to the following generic form of the nonlocal four-fermion interaction

$$S_{\mathrm{int}}^{(4)} = -\frac{G}{2} \sum_\alpha c_\alpha \int \mathrm{d}^4 x \int \mathrm{d}^4 y \, \bar{\psi}(x) \, \Gamma_\alpha \, \psi(y) \, \mathcal{G}(x-y) \, \bar{\psi}(y) \, \Gamma^\alpha \, \psi(x) \,, \tag{3.2.4}$$

with well-defined Fierz expansion coefficients c_α. The Γ_α are a set of Dirac, flavor and color matrices resulting from the Fierz transform, with the property $\gamma_0 \, \Gamma_\alpha^\dagger \, \gamma_0 = \Gamma_\alpha$. By a change of variables, Eq. (3.2.4) can be rewritten as:

$$S_{\mathrm{int}}^{(4)} = -\frac{G}{2} \sum_\alpha c_\alpha \int \mathrm{d}^4 x \int \mathrm{d}^4 z \, \bar{\psi}\left(x+\frac{z}{2}\right) \Gamma_\alpha \, \psi\left(x-\frac{z}{2}\right) \mathcal{G}(z) \, \bar{\psi}\left(x-\frac{z}{2}\right) \Gamma^\alpha \, \psi\left(x+\frac{z}{2}\right). \tag{3.2.4'}$$

The Fierz transformation introduces combinations of operators Γ_α which maintain the symmetries of the original QCD Lagrangian. The symmetry that governs low-energy QCD with two flavors is global chiral SU(2)$_\mathrm{L}$ × SU(2)$_\mathrm{R}$, which undergoes dynamical (spontaneous) breaking to SU(2)$_\mathrm{V}$. Moreover, explicit chiral symmetry breaking corrections are introduced by the quark mass term of the action, $\bar{\psi}\hat{m}\psi$. A minimal subset of operators satisfying this symmetry is the color-singlet pair of scalar-isoscalar and pseudoscalar-isovector operators,

$$\Gamma_\alpha = (1, \mathrm{i}\gamma_5 \, \vec{\tau}) \,, \tag{3.2.5}$$

the one we focus on. Other less relevant operators (vector and axialvector terms in color singlet and color octet channels) will not be treated in this work.

The action constructed up to this point leads to the same gap equation (2.5.8) as obtained from the full QCD Lagrangian using the assumptions of the Dyson-Schwinger approach described in Sect. 2.5.

Separable Four-Fermion Interaction

As it stands, the Dyson-Schwinger gap equation (2.5.8) is an integral equation that is not practical for our present purposes. In particular, the thermodynamics based on this equation would be confronted with the full complexity of finite-temperature Dyson-Schwinger calculations (cf. Refs. [27, 41–44]). We therefore approximate the distribution $\mathcal{G}$ by a separable form, replacing

$$\mathcal{G}(z) \to \int \mathrm{d}^4 z' \, \mathcal{C}\left(z + \frac{z'}{2}\right) \mathcal{C}\left(z - \frac{z'}{2}\right),$$

with $\mathcal{C}(z)$ normalized to $\int \mathrm{d}^4 z \, \mathcal{C}(z) = 1$ (compare also Ref. [45]). The model using this separable ansatz will be worked out starting from the next subsection. A comparison between the full Dyson-Schwinger calculation and the separable approach in Sect. 3.2.4 demonstrates that the separable action is an excellent approximation to the full Dyson-Schwinger formalism for all practical purposes. We can now write the chirally invariant four-fermion interaction in a more tractable form:

$$\mathcal{S}_{\mathrm{int}}^{(4)} = -\frac{G}{2} \int \mathrm{d}^4 x \left[j^S(x) j^S(x) + j_k^P(x) j_k^P(x) \right], \tag{3.2.6}$$

where j_α^S and j_α^P are scalar and pseudoscalar densities, given by[3]

$$
\begin{aligned}
j^S(x) &= \int \mathrm{d}^4 z \, \bar{\psi}\left(x + \frac{z}{2}\right) \mathcal{C}(z) \, \psi\left(x - \frac{z}{2}\right), \\
j_k^P(x) &= \int \mathrm{d}^4 z \, \bar{\psi}\left(x + \frac{z}{2}\right) \mathcal{C}(z) \, \mathrm{i}\gamma_5 \tau_k \psi\left(x - \frac{z}{2}\right).
\end{aligned}
\tag{3.2.7}
$$

The particular functional form of $\mathcal{C}(z)$ used in this work will be given in Sect. 3.2.4.

3.2.2 Formalism

Given the four-fermion coupling in the nonlocal framework, Eq. (3.2.6), we can now write down the complete Euclidean two-flavor nonlocal NJL action, $\mathcal{S}_{\mathrm{E}}$, that will be the basis of the calculations performed in this section. From Eq. (3.2.2) and Eq. (3.2.6) we have

$$\mathcal{S}_{\mathrm{E}} = \int \mathrm{d}^4 x \left\{ \bar{\psi}(x) \left[-\mathrm{i}\gamma^\mu \partial_\mu + \hat{m} \right] \psi(x) - \frac{G}{2} \left[j^S(x) j^S(x) + j_k^P(x) j_k^P(x) \right] \right\}. \tag{3.2.8}$$

In the remainder of this section we demonstrate how this approach works in reproducing zero-temperature QCD, the nonperturbative vacuum and its lowest quark-antiquark excitations: the pseudoscalar pions including their decay constants. As usual, we start from the partition

---

[3] $k \in \{1, 2, 3\}$.

function

$$\mathcal{Z} = \int \mathcal{D}\bar{\psi}\, \mathcal{D}\psi\, e^{-\mathcal{S}_E} \tag{3.2.9}$$

and seek to replace the bilinear combinations of fermionic fields appearing in Eq. (3.2.9) by bosonic fields *(bosonization)*. For this purpose, we introduce four bosonic fields σ and π_k ($k \in \{1,2,3\}$) and four auxiliary fields S, P_k, which serve only for technical reasons in the two-flavor case, but which obtain a more profound significance in the three-flavor generalization. Inserting a "one" in terms of delta functions,

$$1 = \int \mathcal{D}S\, \mathcal{D}P_k\, \delta(S - j^S)\, \delta(P_k - j_k^P) = \int \mathcal{D}S\, \mathcal{D}P_k\, \mathcal{D}\sigma\, \mathcal{D}\pi_k\, e^{\int d^4 z\, \sigma(S - j^S)}\, e^{\int d^4 z\, \pi_k(P_k - j_k^P)},$$

the partition function is written as

$$\mathcal{Z} = \int \mathcal{D}\bar{\psi}\, \mathcal{D}\psi\, \mathcal{D}S\, \mathcal{D}P_k\, \mathcal{D}\sigma\, \mathcal{D}\pi_k\, \exp\left\{ -\int d^4 x\, \left[\bar{\psi}\left(-i\gamma^\mu \partial_\mu + \hat{m} \right)\psi + \sigma j^S + \pi_k j_k^P \right] \right\}$$
$$\times \exp\left\{ \int d^4 x\, \left[(\sigma S + \pi_k P_k) + \frac{G}{2}\left(S^2 + P_k P_k \right) \right] \right\}.$$

The inserted delta functions have implied a replacement of j^S and j_k^P by S and P_k, respectively. The first exponential, on the other hand, contains terms of the form $\bar{\psi}\hat{\mathcal{A}}\psi$, hence the path integration over the fermionic fields $\bar{\psi}$ and ψ can be carried out by standard means, leading to

$$\mathcal{Z} = \int \mathcal{D}\sigma\, \mathcal{D}\pi_k\, \det \hat{\mathcal{A}} \int \mathcal{D}S\, \mathcal{D}P_k\, \exp\left\{ \int d^4 x\, \left[(\sigma S + \pi_k P_k) + \frac{G}{2}\left(S^2 + P_k P_k \right) \right] \right\}. \tag{3.2.10}$$

The integration over the auxiliary fields S, P_k can easily be carried out by completing the square, leading to

$$\mathcal{Z} = \mathcal{N} \int \mathcal{D}\sigma\, \mathcal{D}\pi_k\, \det \hat{\mathcal{A}}\, \exp\left\{ -\frac{1}{2G} \int d^4 x\, \left[\sigma^2 + \vec{\pi}^2 \right] \right\}, \tag{3.2.11}$$

where $\det \hat{\mathcal{A}}$ is the fermion determinant. In momentum space one finds after a simple Fourier transformation[4]

$$\mathcal{A}(p, p') := \langle p | \hat{\mathcal{A}} | p' \rangle = \left(-\not{p} + \hat{m} \right)(2\pi)^4 \delta(p - p') + \mathcal{C}\left(\frac{p + p'}{2} \right)\left[\sigma(p - p') + i\gamma_5 \tau_k \pi_k(p - p') \right]. \tag{3.2.12}$$

Note, that from the normalization of $\mathcal{C}(z)$ in coordinate space we obtain $\mathcal{C}(p = 0) = 1$ in momentum space.

Finally, comparison of Eq. (3.2.11) with the definition (3.2.9) of the partition function gives the bosonized Euclidean action,

$$\mathcal{S}_{\text{bos}} = -\ln \det \hat{\mathcal{A}} + \frac{1}{2G} \int d^4 x\, \left[\sigma^2 + \vec{\pi}^2 \right], \tag{3.2.13}$$

where ln and det denote the functional logarithm and functional determinant, respectively.

---

[4]We use $\hat{\mathcal{C}}(p) = \int d^4 z\, \mathcal{C}(z)\, e^{-ip \cdot z}$, $\hat{\phi}(p) = \int d^4 z\, \phi(z)\, e^{-ip \cdot z}$, where $\phi(z)$ stands for an arbitrary field $\phi \in \{\sigma, \vec{\pi}\}$. We omit the hats on the Fourier transforms and set henceforth $\mathcal{C}(p) \equiv \hat{\mathcal{C}}(p)$, $\phi(p) \equiv \hat{\phi}(p)$, etc.

Mean-Field Approximation and Beyond

In order to perform numerical computations with the bosonized action (3.2.13) we need to expand the full action. This can be accomplished assuming that in the homogeneous and isotropic vacuum, the scalar σ field has a nonzero expectation value $\bar{\sigma} = \langle \sigma \rangle$, while the vacuum expectation values of the pseudoscalar fields π_i are zero. We write $\sigma(x) = \bar{\sigma} + \delta\sigma(x)$, $\vec{\pi}(x) = \delta\vec{\pi}(x)$ and expand the bosonized action (3.2.13) around the mean field in powers of the mesonic fluctuations $\delta\sigma, \delta\vec{\pi}$ (setting $\delta\sigma$ and $\delta\vec{\pi}$ equal to zero defines the so-called *mean-field approximation* (MF)):

$$\mathcal{S}_{\mathrm{E}}^{\mathrm{bos}} = \mathcal{S}_{\mathrm{E}}^{\mathrm{MF}} + \mathcal{S}_{\mathrm{E}}^{(2)} + \dots \tag{3.2.14}$$

The mean-field contribution per four-volume $V^{(4)}$ is given by

$$\frac{\mathcal{S}_{\mathrm{E}}^{\mathrm{MF}}}{V^{(4)}} = -4N_{\mathrm{c}} \int \frac{\mathrm{d}^4 p}{(2\pi)^4} \ln\left[p^2 + M^2(p)\right] + \frac{\bar{\sigma}^2}{2G} , \tag{3.2.15}$$

with the mass function $M(p)$ determined by the gap equation

$$M(p) = m_0 + \mathcal{C}(p)\,\bar{\sigma} . \tag{3.2.15a}$$

Here m_0 stands for either the current up- or down-quark masses, as we are assuming isospin symmetry. The quadratic terms beyond mean-field approximation are derived explicitly in Appendix C. Here we only state the result:

$$\mathcal{S}_{\mathrm{E}}^{(2)} = \frac{1}{2} \int \frac{\mathrm{d}^4 p}{(2\pi)^4} \left[F^+(p^2)\,\delta\sigma(p)\,\delta\sigma(-p) + F^-(p^2)\,\delta\vec{\pi}(p) \cdot \delta\vec{\pi}(-p) \right], \tag{3.2.16}$$

where

$$F^{\pm}(p^2) = \frac{1}{G} - 8N_{\mathrm{c}} \int \frac{\mathrm{d}^4 q}{(2\pi)^4}\, \mathcal{C}^2(q) \frac{q^+ \cdot q^- \mp M(q^+)M(q^-)}{\left[(q^+)^2 + M^2(q^+)\right]\left[(q^-)^2 + M^2(q^-)\right]}$$

$$= \frac{1}{G} - \Gamma^{\pm} \underset{q+\frac{p}{2}}{\overset{q-\frac{p}{2}}{\bigcirc}} \Gamma^{\pm} \tag{3.2.17}$$

with $q^{\pm} = q \pm \frac{p}{2}$ and Γ^+, Γ^- denote the scalar-isoscalar and pseudoscalar-isovector vertices, respectively. The loop diagram involves the fermion (quark) quasiparticle propagator (2.5.9), already met when dealing with Dyson-Schwinger equations in Sect. 2.5,

$$S_{\mathrm{F}}(p) = \underline{\qquad\qquad} = \frac{1}{-\not{p} + M(p)} , \tag{2.5.9}$$

with the momentum-dependent (dynamical) constituent quark mass $M(p)$.

Gap Equation and Chiral Condensate

The mean-field part, $\mathcal{S}_{\mathrm{E}}^{\mathrm{MF}}$, of the action $\mathcal{S}_{\mathrm{E}}$ is governed by the scalar mean field $\bar{\sigma}$. Its value is found by the principle of least action, $\frac{\delta \mathcal{S}_{\mathrm{E}}^{\mathrm{MF}}}{\delta \sigma} = 0$ for $\sigma = \bar{\sigma}$. It is straightforward to derive

$$\bar{\sigma} = 8 N_{\mathrm{c}} G \int \frac{\mathrm{d}^4 p}{(2\pi)^4} \, \mathcal{C}(p) \frac{M(p)}{p^2 + M^2(p)} \,, \qquad (3.2.18)$$

to be solved self-consistently with the gap equation (3.2.15a). We notice, that Eq. (3.2.18) is the same as the Dyson-Schwinger gap equation (2.5.12) provided one uses the separable ansatz $\mathcal{G}(p-q) = \mathcal{C}(p)\mathcal{C}(q)$ in Eq. (2.5.12) and sets $M(p) = m_0 + \bar{\sigma} \, \mathcal{C}(p)$. The assumption of separability turns out to be only a weak restriction, but it reduces drastically the technical complexity: while Eq. (2.5.12) is an integral equation for which the self-consistent solution $M(p)$ has to be known over the whole momentum region, Eq. (3.2.18) represents an equation for the mean-field value $\bar{\sigma}$, thence only for a single number. This simplification will allow one to extend the formalism presented in this chapter to QCD thermodynamics (Chapt. 5).

The chiral condensate $\langle \bar{\psi}\psi \rangle = \langle \bar{u}u \rangle + \langle \bar{d}d \rangle$ can be calculated from $\mathcal{S}_{\mathrm{E}}^{\mathrm{bos}}$ using the Feynman-Hellmann theorem, by differentiation with respect to the current quark mass m_0. Equivalently, we can use the definition (2.4.9)

$$\langle \bar{\psi}\psi \rangle = -\mathrm{i} \, \mathrm{Tr} \lim_{y \to x^+} \left[S_{\mathrm{F}}(x,y) - S_{\mathrm{F}}^{(0)}(x,y) \right], \qquad (2.4.9)$$

with the full fermion Green function,

$$S_{\mathrm{F}}(x,y) = \int \frac{\mathrm{d}^4 p}{(2\pi)^4} \, e^{\mathrm{i}p(x-y)} S_{\mathrm{F}}(p) \,, \qquad (3.2.19)$$

and the free quark propagator $S_{\mathrm{F}}^{(0)}$ subtracted. Using Eq. (2.5.9) with $M(p)$ from Eq. (3.2.15a) this leads to

$$\langle \bar{\psi}\psi \rangle = -4 N_{\mathrm{f}} N_{\mathrm{c}} \int \frac{\mathrm{d}^4 p}{(2\pi)^4} \left[\frac{M(p)}{p^2 + M^2(p)} - \frac{m_0}{p^2 + m_0^2} \right]. \qquad (3.2.20)$$

Note that $M(p) \to m_0$ for large p. The subtraction makes sure that no perturbative artifacts are left over in $\langle \bar{\psi}\psi \rangle$ for $m_0 \neq 0$.

We have already commented on the parallels and differences of the nonlocal NJL model and the more general Dyson-Schwinger formalism; a more thorough investigation will be presented in Sect. 3.2.4. Before continuing with the formal derivation of all quantities of interest, we will first examine a comparison with the local NJL model. As described in Sect. 3.1, $\mathcal{C}(p)$ and $M(p)$ are replaced by constants in the local NJL model, $\mathcal{C}(0) = 1$ and M. The relevant momentum-space integrations are cut off at $|\vec{p}| = \Lambda_{\mathrm{NJL}}$. The gap equation is simply $M = -G\langle \bar{\psi}\psi \rangle = \bar{\sigma}$. This direct proportionality between the scalar mean field $\bar{\sigma}$ and the chiral condensate is not realized anymore in the nonlocal model, as the inspection of Eqs. (3.2.18) and (3.2.20) demonstrates.

Pion Mass, Quark-Pion Coupling Constant and Pion Decay Constant

The pion mass m_π is determined by the pole of the pion propagator, while the square of the quark-pion coupling constant, $g_{\pi qq}^2$, figures as the residue of the pion pole in the pseudoscalar-isovector quark-antiquark amplitude,

$$D_{ij}^\pi(q) = \mathrm{i}\gamma_5 \tau_i \frac{\mathrm{i}g_{\pi qq}^2}{q^2 + m_\pi^2} \mathrm{i}\gamma_5 \tau_j \, . \tag{3.2.21}$$

Its inverse is easily calculated from the bosonized action (3.2.13) by means of functional differentiation, i.e.,

$$\left[D_{ij}^\pi(q)\right]^{-1} = \frac{\delta^2 \mathcal{S}_{\mathrm{E}}^{\mathrm{bos}}}{\delta \pi_i(q)\delta \pi_j(0)} = \delta_{ij} F^-(q^2) \, . \tag{3.2.22}$$

The last equality results from Eqs. (3.2.16) and (3.2.17). The pion mass is therefore determined by

$$F^-(-m_\pi^2) = 0 \, . \tag{3.2.23}$$

Furthermore, by comparison with Eq. (3.2.21) the quark-pion coupling constant is given as:

$$g_{\pi qq}^{-2} = \left. \frac{\mathrm{d}F^-(q^2)}{\mathrm{d}q^2} \right|_{q^2 = -m_\pi^2} \, . \tag{3.2.24}$$

The calculation of the pion decay constant f_π is more involved. As we have seen in Sect. 2.4.2, it is defined through the matrix element of the axial current $J_{\mathrm{A},i}^\mu(x)$ between the vacuum and the physical one-pion state $\tilde{\pi}_j$ with $\tilde{\pi}_j = g_{\pi qq}^{-1} \pi_j$ (cf. Eq. (2.4.10)):

$$\langle 0|J_{\mathrm{A},i}^\mu(0)|\tilde{\pi}_j(p)\rangle = \mathrm{i}\delta_{ij} p^\mu f_\pi \, . \tag{3.2.25}$$

Note that the usual local axial current, $J_{\mathrm{A},i}^{\mu,\mathrm{loc}} = \bar{\psi}(x)\gamma_5 \frac{\tau_i}{2}\gamma^\mu \psi(x)$, is not conserved within the nonlocal framework (cf. Ref. [46, 47]). In order to calculate this matrix element one has to gauge the nonlocal action in Eq. (3.2.13) following the procedure outlined in Sects. 2.2.1 and 2.4.3. This requires not only the replacement of the partial derivative by a covariant derivative,

$$\partial_\mu \to \partial_\mu + \frac{\mathrm{i}}{2}\gamma_5 \tau_i \, \mathcal{A}_\mu^i(x),$$

where $\mathcal{A}_\mu^i$ ($i \in \{1, 2, 3\}$) are a set of axial gauge fields, but also the connection of nonlocal terms through a parallel transport with a Wilson line (cf. Eq. (2.4.15) and Refs. [48, 49]),

$$\mathcal{W}(x,y) = \mathcal{P}\exp\left\{\frac{\mathrm{i}}{2}\int_0^1 \mathrm{d}\alpha \, \gamma_5 \tau_i \, \mathcal{A}_i^\mu(x + (y-x)\alpha)\,(y_\mu - x_\mu)\right\},$$

where we have chosen a straight line connecting the points x and y. This means that expressions of the form $\bar{\psi}(x)\hat{\mathcal{O}}(z)\psi(y)$ (where $\hat{\mathcal{O}}(z)$ is an arbitrary field operator) in the action $\mathcal{S}_{\mathrm{E}}$, Eq. (3.2.8), have to be replaced by $\bar{\psi}(x)\,\mathcal{W}(x,z)\,\hat{\mathcal{O}}(z)\,\mathcal{W}(z,y)\,\psi(y)$. This guarantees the (local) gauge invariance of the underlying Lagrangian. It turns out that the only term that is eventually affected by the gauging is the fermion determinant of $\hat{\mathscr{A}}$, Eq. (3.2.12), which then becomes, in coordinate space,

$$
\mathscr{A}^{G}(x,y) = \left(-i\,\partial_y + \frac{1}{2}\gamma_5\tau_i\,\mathscr{A}^i + \hat{m}\right)\delta(x-y) +
$$
$$
+\, \mathcal{C}(x-y)\,\mathcal{W}\left(x,\frac{x+y}{2}\right)\Gamma_i\,\phi_i\left(\frac{x+y}{2}\right)\mathcal{W}\left(\frac{x+y}{2},y\right). \tag{3.2.26}
$$

Here Γ_i stands either for $\Gamma = 1$ or $\Gamma_i = i\,\gamma_5\tau_i$, and ϕ_i accordingly for either the scalar field σ, or a pseudoscalar pion field, π_i. The desired matrix element then follows from the gauged fermion determinant according to

$$
\langle 0|J^{\mu}_{A,i}(0)|\pi_j(p)\rangle = -\left.\frac{\delta^2 \ln \det \mathscr{A}^{G}}{\delta\pi_j(p)\,\delta\mathcal{A}^i_{\mu}(t)}\right|_{\substack{A=0 \\ t=0}}. \tag{3.2.27}
$$

After a lengthy evaluation (all details of this calculation are presented in Appendix D) of the functional derivatives we obtain in momentum space[5]

$$
\langle 0|J^{\mu}_{A,i}(0)|\pi_j(p)\rangle = 2i\,\mathrm{tr}\{\tau_i,\tau_j\}\,\widetilde{\mathrm{Tr}}\left\{\int_0^1 d\alpha\, q_\mu \frac{d\mathcal{C}(q)}{dq^2}\frac{M(q^+_\alpha)}{q^{+2}_\alpha + M^2(q^+_\alpha)}\right\} +
$$
$$
+\, 2i\,\mathrm{tr}\{\tau_i,\tau_j\}\,\widetilde{\mathrm{Tr}}\left\{\mathcal{C}(q)\frac{q^-_\mu M(q^-)}{\left(q^{+2}+M^2(q^+)\right)\left(q^{-2}+M^2(q^-)\right)}\right\} +
$$
$$
+\, 2i\,\bar{\sigma}\,\mathrm{tr}\{\tau_i,\tau_j\}\times
$$
$$
\times\,\widetilde{\mathrm{Tr}}\left\{\int_0^1 d\alpha\, q_\mu \frac{d\mathcal{C}(q)}{dq^2}\,\mathcal{C}\left(q-\frac{p}{2}\alpha\right)\frac{q^+_\alpha \cdot q^-_\alpha + M(q^+_\alpha)M(q^-_\alpha)}{\left(q^{+2}_\alpha+M^2(q^+_\alpha)\right)\left(q^{-2}_\alpha+M^2(q^-_\alpha)\right)}\right\}, \tag{3.2.28}
$$

where $\widetilde{\mathrm{Tr}}$ stands for the functional trace and, moreover,

$$
q^+_\alpha = q + \frac{p}{2}(1-\alpha), \qquad q^-_\alpha = q - \frac{p}{2}(1+\alpha)
$$
$$
q^+ = q + \frac{p}{2}, \qquad\qquad q^- = q - \frac{p}{2}. \tag{3.2.29}
$$

Now, the pion decay constant can be derived from its definition, Eq. (3.2.25) and the expression (3.2.28), by contraction with p^μ:

$$
f_\pi = i\,p_\mu\langle 0|J^{\mu}_{A,i}(0)|\pi_i(p)\rangle\frac{g^{-1}_{\pi qq}}{m^2_\pi}, \tag{3.2.30}
$$

evaluated at the pion pole $p^2 = -m^2_\pi$. One pion-field component with index i is singled out in this expression.

Chiral Relations

In order to confirm that this nonlocal model is consistent with fundamental chiral symmetry requirements, we derive the Goldberger-Treiman and Gell-Mann–Oakes–Renner relations. For

---

[5]$\mathrm{tr}\{\tau_i,\tau_j\} = 4\delta_{ij}$ can readily be evaluated. We write the result in this more general form, though, because this will enable us the generalize the matrix element (3.2.28) immediately to three flavors.

this purpose it is sufficient to consider an expansion of Eq. (3.2.30) up to $\mathcal{O}(m_\pi^2)$, i.e., to leading (linear) order in the quark mass m_0, giving

$$f_\pi = g_{\pi qq} \frac{\mathcal{F}(-m_\pi^2) - \mathcal{F}(0)}{m_\pi^2} \,, \qquad (3.2.30')$$

where we have defined

$$\mathcal{F}(p^2) = m_0 \, \mathcal{I}_1(p^2) + \bar{\sigma} \, \mathcal{I}_2(p^2) \,,$$

with

$$\mathcal{I}_1 = 8N_c \int \frac{d^4 q}{(2\pi)^4} \, \mathcal{C}(q) \frac{q \cdot (q+p) + M(q)M(q+p)}{[q^2 + M^2(q)]\,[(q+p)^2 + M^2(q+p)]}$$

$$\mathcal{I}_2 = 8N_c \int \frac{d^4 q}{(2\pi)^4} \, \mathcal{C}(q)\,\mathcal{C}(q+p) \frac{q \cdot (q+p) + M(q)M(q+p)}{[q^2 + M^2(q)]\,[(q+p)^2 + M^2(q+p)]} \,.$$

Using Eqs. (3.2.18) and (3.2.23), one can write $\mathcal{F}(0) = \bar{\sigma}/G$ and $\mathcal{I}_2(-m_\pi^2) = 1/G$. Then one obtains from Eq. (3.2.30')

$$m_\pi^2 f_\pi = m_0 \, g_{\pi qq} \, \mathcal{I}_1(-m_\pi^2) \,. \qquad (3.2.31)$$

A Taylor expansion around the chiral limit of the pion polarization term, Eq. (3.2.17), leads to

$$F^-(p^2) = F^-(0) + \left.\frac{\partial F^-}{\partial p^2}\right|_{p^2=0} p^2 + \mathcal{O}(p^4) = F^-(0) + g_{\pi qq}^{-2} p^2 + \mathcal{O}(p^4) \,.$$

Using this in the conditional equation for the pion mass, $F^-(-m_\pi^2) = 0$, one recovers the leading order contribution to m_π^2:

$$m_\pi^2 = \frac{g_{\pi qq}^2}{\bar{\sigma}} m_0 \mathcal{I}_1 + \mathcal{O}(m_\pi^4) = g_{\pi qq}^2 \, F^-(0) + \mathcal{O}(m_\pi^4) \,. \qquad (3.2.32)$$

Finally, combining Eqs. (3.2.17) and (3.2.20) gives

$$F^-(0) = -\frac{m_0}{\bar{\sigma}^2} \langle \bar{\psi}\psi \rangle + \mathcal{O}(m_0^2) \,.$$

From Eq. (3.2.31) and the first equality of Eq. (3.2.32) one finds

$$f_\pi \, g_{\pi qq} = \bar{\sigma} + \mathcal{O}(m_0) \,. \qquad (3.2.33)$$

In the chiral limit, i.e., for $m_0 = 0$, this is the Goldberger-Treiman (compare Eq. (2.4.13)) relation at the level of quarks as quasiparticles. Using this relation, the second equality of Eq. (3.2.32) together with the expression for $F^-(0)$ derived above gives

$$m_\pi^2 f_\pi^2 = -m_0 \langle \bar{\psi}\psi \rangle + \mathcal{O}(m_0^2) \,, \qquad (3.2.34)$$

which is the well-known Gell-Mann–Oakes–Renner relation (compare Eq. (2.4.12)). We have thus demonstrated that the nonlocal NJL model preserves chiral low-energy theorems and current-algebra relations.

3.2.3 Fixing of the Distribution $\mathcal{C}(p)$: Politzer's Quark Self-Energy

The contact strength G will be determined by fixing pion properties (mass and decay constant). This requires choosing a particular functional form for the momentum distribution function $\mathcal{C}(p)$. The gap equation (3.2.18) of the local NJL model implies a simple relation between the dynamical (constituent) quark mass and the chiral condensate, $M = -G\langle\bar{\psi}\psi\rangle$. A similar expression, although not completely equivalent, holds in the nonlocal model, compare Eqs. (3.2.18) and (3.2.20): $M(p) = \mathcal{C}(p)\bar{\sigma}$ (in the chiral limit). It is this relation that will enable us to fix $\mathcal{C}(p)$ and give this distribution a physical foundation.

We consider a quark propagating with large (Euclidean) momentum p in the QCD vacuum. Its self-energy $\Sigma(p)$ is given pictorially as

$$-\mathrm{i}\Sigma(p) = \quad\text{}\quad , \tag{3.2.35}$$

with full quark and gluon propagators and vertex functions (cf. Sect. 2.5). At high momentum, this self-energy can be evaluated recalling its operator product expansion and identifying the leading $\mathcal{O}(p^{-2})$ term [50, 51].

It is well known that QCD perturbation theory at any order does not generate a nonzero quark mass term starting from a massless quark, simply as a consequence of helicity conservation at each quark-gluon vertex. However, as pointed out by Politzer [50], the presence of a nonperturbative QCD vacuum with a quark condensate $\langle\bar{\psi}\psi\rangle \neq 0$ turns a massless current quark into a quasiparticle with a momentum-dependent mass (often referred to as a constituent quark mass): the spontaneous breaking of chiral symmetry implies a nonzero dynamical quark mass.

At high quark momentum, the leading nontrivial part of $\Sigma(p)$, calculated in "rainbow truncation" and "Abelian approximation" (see Sect. 2.5) is found, with the running QCD coupling $\alpha_{\mathrm{s}}(p^2)$, Eq. (2.3.8), to leading order in $1/p^2$ (with $p^2 > 0$):

$$\Sigma(p^2) = \pi\frac{N_c^2 - 1}{2N_c^2 N_{\mathrm{f}}}\frac{\alpha_{\mathrm{s}}(p^2)}{p^2}(3 + \xi)\langle\bar{\psi}\psi\rangle + \delta\Sigma , \tag{3.2.36}$$

where we have explicitly written the nonperturbative piece, $\Sigma^{\mathrm{n.\,p.}}$, and the remaining $\delta\Sigma$ stands for all perturbative corrections to Σ (which vanish in the chiral limit, $m_0 \to 0$). The parameter ξ is required for gauge fixing (see also Refs. [51, 52]).

From Eq. (3.2.36) one arrives at the dynamically generated quark mass term

$$M(p^2) := \Sigma^{\mathrm{n.\,p.}}(p^2) = -\pi\frac{N_c^2 - 1}{2N_c^2 N_{\mathrm{f}}}\frac{\alpha_{\mathrm{s}}(p^2)}{p^2}(3 + \xi)\langle\bar{\psi}\psi\rangle . \tag{3.2.37}$$

This is the result for the constituent quark mass calculated first by Politzer in an operator product expansion. Even in the limit of vanishing current quark masses, the quark propagator has a nonperturbative contribution which generates a large constituent quark mass in the presence of a nonvanishing chiral (quark) condensate. We choose the Landau gauge, setting $\xi = 0$, as in Sect. 2.5. Any other value of ξ can be absorbed by a rescaled coupling in Eq. (3.2.37).

From Eq. (3.2.15a) it is evident that the distribution $\mathcal{C}(p)$ must behave as

$$\mathcal{C}(p^2) \propto \frac{2\pi}{3} \frac{\alpha_s(p^2)}{p^2} \qquad (p \geq \Gamma) , \tag{3.2.38}$$

at p larger than a matching scale Γ of order $1\,\text{GeV}$ below which $\mathcal{C}(p^2)$ is governed by nonperturbative physics. For our present purposes it is sufficient to use the standard form for the perturbative running QCD coupling,

$$\alpha_s(p^2) = \frac{4\pi}{\beta_0 \ln \frac{p^2}{\Lambda^2_{\text{QCD}}}} , \tag{2.3.8}$$

with $\beta_0 = 9$ (using three active quark flavors) and $\Lambda_{\text{QCD}} = 0.25\,\text{GeV}$ in order to reproduce $\alpha_s = 0.12$ at $m_Z = 91.2\,\text{GeV}$.

The low-momentum behavior of $\mathcal{C}(p)$ is governed by nonperturbative QCD effects. It is clear that the expression (3.2.38) breaks down at some $p \lesssim 1\,\text{GeV}$ because of the Landau pole in Eq. (2.3.8). Here we choose a Gaussian $\mathcal{C}(p) = \exp\left[-\frac{p^2 d^2}{2}\right]$ for convenience in order to match Eq. (3.2.38) below Γ. Owing to the integral measure $\mathrm{d}^4 p \sim p^3 \,\mathrm{d}^3 p$ in the gap equation (3.2.18), the low-momentum behavior of $\mathcal{C}(p)$ is strongly suppressed. Our results are, therefore, only weakly influenced by the particular choice of $\mathcal{C}(p)$ at $p \ll \Gamma$. A more profound physical justification for this choice can be given from the instanton liquid model, dealt with in Sect. 3.3.1. From this the width of the Gaussian d is identified with the instanton size, $d \simeq 1/3\,\text{fm}$. Putting these two assumptions together, we have the nonlocality distribution

$$\mathcal{C}(p^2) = \begin{cases} e^{-p^2 d^2/2} & \text{for } p^2 < \Gamma^2 \\ \text{const.} \cdot \dfrac{\alpha_s(p^2)}{p^2} & \text{for } p^2 \geq \Gamma^2 , \end{cases} \tag{3.2.39}$$

normalized as $\mathcal{C}(p = 0) = 1$ and with a constant fixed by the matching condition requiring continuity and differentiability of $\mathcal{C}(p)$ at $p = \Gamma$. The full momentum distribution function $\mathcal{C}(p)$ used for the $N_f = 2$ flavor model is shown in Fig. 3.1.

3.2.4 Parameters

Given the nonlocality distribution function $\mathcal{C}(p)$ (3.2.39) and the formulas for the pion mass and decay constant in Sect. 3.2.2, we can know fix the parameters of the two-flavor nonlocal NJL model. There are actually only two parameters apart from the matching point Γ: the four-fermion interaction strength G and the current quark mass m_0. The width d of the Gaussian is correlated with the matching parameter Γ by the matching condition for $\mathcal{C}(p)$ at $p = \Gamma$.

The parameters and resulting values of observables are listed in Table 3.1. The determination of G, Γ and m_0 is such as to reproduce, as closely as possible, the empirical pion decay constant $f_\pi = 92.4\,\text{MeV}$ and the pion mass m_π. Fig. 3.2 shows the momentum dependence of the dynamical quark mass $M(p)$ compared with lattice data from Ref. [53]. One notes that the nonlocal NJL model reproduces both the low- and the high-momentum behavior. The quark mass

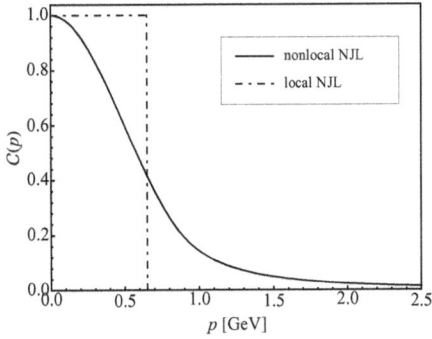

FIGURE 3.1: Distribution $\mathcal{C}(p)$ used in the non-local NJL model (solid line). The dot-dashed line shows $\mathcal{C}(p)$ in a local NJL model with four-momentum cutoff.

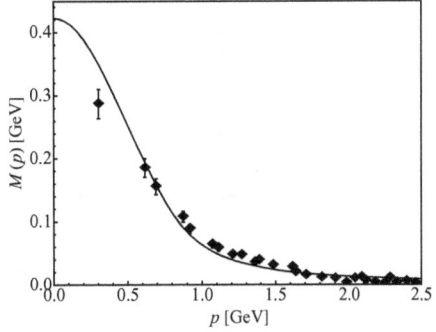

FIGURE 3.2: Momentum dependence of the constituent quark mass M, calculated in the two-flavor nonlocal NJL model, compared with lattice data in Landau gauge extrapolated to the chiral limit (from Ref. [53], improved staggered quark actions have been used).

m_0	Γ	G	$\langle \bar{\psi}\psi \rangle^{1/3}$	$\bar{\sigma}$	f_π	m_π
$3.3\,\mathrm{MeV}$	$(0.24\,\mathrm{fm})^{-1}$	$(1.26\,\mathrm{fm})^2$	$-0.36\,\mathrm{GeV}$	$0.42\,\mathrm{GeV}$	$0.09\,\mathrm{GeV}$	$0.14\,\mathrm{GeV}$

TABLE 3.1: Parameters and calculated physical quantities for the nonlocal NJL model with $N_\mathrm{c} = 3$ and $N_\mathrm{f} = 2$.

$m_0 \simeq 3.3\,\mathrm{MeV}$ is compatible with QCD estimates at a typical renormalization scale $\mu \simeq 2\,\mathrm{GeV}$ [19]. The value of the chiral condensate follows consistently from the Gell-Mann–Oakes–Renner relation (2.4.12), with $\langle \bar{u}u \rangle = \langle \bar{d}d \rangle \simeq -(0.28\,\mathrm{GeV})^3$.

The optimal matching scale $\Gamma \simeq 0.83\,\mathrm{GeV} \simeq (0.24\,\mathrm{fm})^{-1}$ corresponds to $d \simeq 0.4\,\mathrm{fm}$ which is indeed compatible with the typical instanton size $d_\mathrm{I} \simeq 1/3\,\mathrm{fm}$, cf. Sect. 3.3.1. It turns out, however, that Γ is located in the momentum window where the gap equation (3.2.18) has its maximum weight. It is therefore important to examine the sensitivity of the results with respect to variations of Γ. We have performed this test by varying the matching scale in the range $0.6\,\mathrm{GeV} < \Gamma < 1\,\mathrm{GeV}$ (i. e., about $20\,\%$ around its optimal value), with the constraint of keeping the chiral condensate $\langle \bar{\psi}\psi \rangle$ fixed. The resulting variations in the pion mass and decay constant are within only $10\,\%$, implying stable conditions.

In order to understand the meaning of the coupling strength G we note that a direct comparison to the NJL model is not appropriate as the NJL model uses a three-momentum cutoff for the regularization of loop integrals while in the nonlocal approach the distribution $\mathcal{C}(p)$ mimics a four-momentum cutoff. This implies, in particular, a different scale for the coupling strength G appearing in the scalar mean field $\bar{\sigma}$. Consider again the chiral limit and compare Eqs. (3.2.18) and (3.2.20). Writing Eq. (3.2.15a) as $M(0) = m_0 - \frac{\bar{\sigma}}{-\langle \bar{\psi}\psi \rangle} \langle \bar{\psi}\psi \rangle$ and comparing this with the constituent quark mass in the local NJL model, $M_\mathrm{NJL} = m_0 - G\langle \bar{\psi}\psi \rangle$, we can interpret

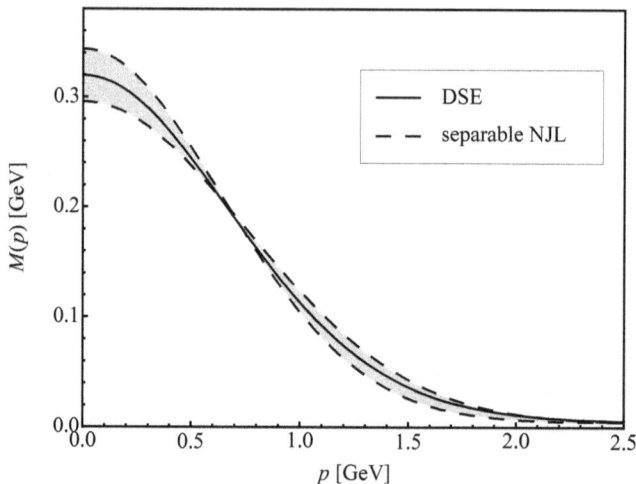

FIGURE 3.3: Solution to the DSE gap equation Eq. (2.5.12) (solid line) compared to the solution of the separable ansatz presented in this section, Eq. (3.2.18) (dashed curves) using a Gaussian ansatz for the functions $\mathcal{G}(p)$ and $\mathcal{C}(p)$, respectively. The gray band indicates different values for the chiral condensate, varying between $(260\,\mathrm{MeV})^3 \leq |\langle \bar{u}u \rangle| = |\langle \bar{d}d \rangle| \leq (280\,\mathrm{MeV})^3$; for the DSE we obtained $|\langle \bar{u}u \rangle| = (280\,\mathrm{MeV})^3$.

$\tilde{G} := -\frac{\bar{\sigma}}{\langle \bar{\psi}\psi \rangle}$ as an effective coupling strength in a corresponding local NJL model. Indeed, one obtains $\tilde{G} \approx 9\,\mathrm{GeV}^{-2}$ which is very close to the usual local NJL value found in the literature (e. g., Refs. [35, 37, 38, 54]).

We conclude this section summarizing the main features of the nonlocal NJL model. The prime advantage of the present nonlocal approach is the absence of an artificial momentum-space cutoff as it appears in local NJL-type models. Indeed, the gap equation (3.2.18) does not require any regularization: the scalar mean field $\bar{\sigma}$ and, likewise, the dynamical quark mass $M(p)$ are well-defined within the nonlocal scheme.[6] As a further check, we compare this separable model with the full Dyson-Schwinger gap equation (2.5.8) and its solution (Fig. 2.1). For this purpose we solve the separable model of this section using a Gaussian distribution function $\mathcal{C}(p)$ in order to be consistent with the numerical calculation presented in Sect. 2.5. After fixing the parameters such as to reproduce the experimental values of the pion mass and decay constant we obtain the results shown in Fig. 3.3. This plot clearly demonstrates that the impact of the separability assumption on the gluon correlation function, i. e., $\mathcal{G}(p-q) \to \mathcal{C}(p) \cdot \mathcal{C}(q)$ is very weak and can basically be absorbed by a redefinition of the model parameters.

---

[6]This does not exclude the possibility that secondary quantities, such as the chiral condensate, can be weakly divergent and potentially require a cutoff at ultrahigh momenta, very far beyond the range of applicability of the model. Inspection of the integral for $\langle \bar{\psi}\psi \rangle$ in Eq. (3.2.20), with the asymptotic form $M(p) \to m_q + \frac{\mathrm{const.}}{p^2 \ln p^2}$ inserted, displays a weak (double-logarithmic) far-ultraviolet divergence. For regularization we choose a cutoff at $20\,\mathrm{GeV}$, the necessity of which just reflects the simplified leading-order choice for the asymptotics of $\mathcal{C}(p)$.

3.3 Three-Flavor Nonlocal NJL Model:
Role of Strangeness and Axial Anomaly

This section extends the nonlocal NJL model to three quark flavors [40]. It is straightforward to promote the chirally invariant four-fermion interaction $\mathcal{S}_{\text{int}}^{(4)}$, Eq. (3.2.6), to three flavors, replacing the Pauli matrices by the generators of the $N_{\text{f}} = 3$ flavor group, with the Gell-Mann matrices $\lambda_i, i \in \{1, \ldots, 8\}$ and the singlet matrix[7] $\lambda_0 := \sqrt{\frac{2}{3}}\text{diag}(1,1,1)$. For completeness, we recall

$$\mathcal{S}_{\text{int}}^{(4)} = -\frac{G}{2} \int \mathrm{d}^4x \left[j_\alpha^S(x) j_\alpha^S(x) + j_\alpha^P(x) j_\alpha^P(x) \right], \tag{3.3.1}$$

where j_α^S and j_α^P are scalar and pseudoscalar densities, given by

$$\begin{aligned}
j_\alpha^S(x) &= \int \mathrm{d}^4z\, \bar{\psi}\left(x + \frac{z}{2}\right) \mathcal{C}(z) \lambda_\alpha \psi\left(x - \frac{z}{2}\right), \\
j_\alpha^P(x) &= \int \mathrm{d}^4z\, \bar{\psi}\left(x + \frac{z}{2}\right) \mathcal{C}(z) i\gamma_5 \lambda_\alpha \psi\left(x - \frac{z}{2}\right).
\end{aligned} \tag{3.3.2}$$

The functional form of $\mathcal{C}$ in momentum space is again given by Eq. (3.2.39).

3.3.1 The QCD Vacuum (II): Axial Anomaly and Instantons

It is a much more involved task to derive the term that introduces the anomalous $\text{U}(1)_\text{A}$ breaking. In this subsection we show how the axial $\text{U}(1)$ anomaly arises in QCD and give an explanation guided by instantons. The instanton model leads to a nonlocal separable six-fermion interaction expression that can immediately be transferred to the nonlocal generalization of the three-flavor local NJL Lagrangian (3.1.2).

Anomalous Breaking of the Axial Symmetry

The QCD Lagrangian is invariant under $\text{U}(1)_\text{A}$ transformations of the fermion fields $\psi, \bar{\psi}$ in the chiral limit. At the classical level, the flavor-singlet axial-vector current $j_5^\mu(x) = \bar{\psi}\gamma^\mu\gamma_5\psi$ is conserved, $\partial_\mu j_5^\mu(x) = 0$. This statement does no longer hold when deriving the continuity equation in the quantum field theory. The appearance of an anomalous term in the divergence of the axial-vector current was first shown by Adler [55] and Bell and Jackiw [56] in 1969.

We outline here the more general proof of K. Fujikawa [57] using the functional integral formalism; this proof holds to all orders of perturbation theory. The starting point is the derivation of the axial-vector Ward identities form the functional integral (compare Eq. (2.3.3')[8])

$$\mathcal{Z} = \int \mathcal{D}A\, \mathcal{D}\psi\, \mathcal{D}\bar{\psi} \exp\left\{ -\int \mathrm{d}^4x\, \mathscr{L}_{\text{E}}^{\text{QCD}} \right\} \tag{3.3.3}$$

with the Lagrangian and action defined as in Eq. (2.5.6). Next, one considers an infinitesimal

---

[7]The normalization has been chosen, as in the two-flavor case, such that $\text{tr}\{\lambda_\alpha \cdot \lambda_\beta\} = 2\delta_{\alpha\beta}$ holds for all $\alpha, \beta \in \{0, \ldots, 8\}$.

[8]The source terms in Eq. (2.3.3') are suppressed here.

axial U(1) transformation on the fermion fields

$$\psi(x) \to \psi'(x) = \left(1 + i\alpha(x)\gamma^5\right)\psi(x)$$
$$\bar{\psi}(x) \to \bar{\psi}'(x) = \bar{\psi}\left(1 + i\alpha(x)\gamma^5\right)$$

(3.3.4)

with a real function $\alpha(x)$. This transformation causes a nontrivial Jacobian $\mathcal{J}$ of the (fermionic) path integral measure, i. e., one obtains [57] for N_f flavors

$$\mathcal{J} = \exp\left\{\int \mathrm{d}^4x\,\alpha(x)\left[\frac{g^2 N_f}{32\pi^2}\varepsilon^{\mu\nu\lambda\sigma}G_{\mu\nu}G_{\lambda\sigma}\right]\right\}$$

(3.3.5)

with the field strength tensor (2.2.2). The tensor

$$\tilde{G}^{\mu\nu} := \frac{1}{2}\varepsilon^{\mu\nu\lambda\sigma}G_{\lambda\sigma}$$

(3.3.6)

is the *dual* field strength tensor. The functional integral can then be written as

$$\mathcal{Z} \xrightarrow[\mathrm{U(1)_A}]{} \int \mathcal{D}A\,\mathcal{D}\psi\,\mathcal{D}\bar{\psi}\,\exp\left\{-\int \mathrm{d}^4x\left[\mathscr{L}_E^{\mathrm{QCD}} + \alpha(x)\left(\partial_\mu j_5^\mu + \frac{g^2 N_f}{32\pi^2}\varepsilon^{\mu\nu\lambda\sigma}G_{\mu\nu}G_{\lambda\sigma}\right)\right]\right\},$$

(3.3.7)

where the term $\alpha(x)\partial_\mu j_5^\mu$ is a result of the axial transformation in the kinetic part of the fermionic action. The Ward identity associated with the symmetry is given by the derivative of the functional integral with respect to $\alpha(x)$. It follows that

$$\partial_\mu j_5^\mu = -\frac{g^2 N_f}{32\pi^2}\varepsilon^{\mu\nu\lambda\sigma}G_{\mu\nu}^a G_{\lambda\sigma}^a\,.$$

(3.3.8)

Therefore, at the quantum level, the axial-vector current is not even conserved in the chiral limit. After this mathematical derivation of the axial anomaly, we give in the next subsection a physical interpretation of the anomalous breaking of the $U(1)_A$ symmetry.

Instantons

The energy density of the QCD vacuum is $\epsilon_0 \simeq -500\,\mathrm{MeV/fm^3}$. This quantity is related through the energy-momentum tensor to the quark and gluon condensate and hence to chiral symmetry breaking and confinement. A model of the QCD vacuum should therefore be able to explain the origin of this value. A frequently discussed option is the instanton model. Instantons are classical solutions to the Euclidean equations of motion. They are characterized by a topological quantum number and correspond to tunneling events between degenerate classical vacua. It is these tunneling events that lower the ground-state energy and give a possible explanation for the negative nonperturbative vacuum energy density. We will see later that instantons are associated with fermionic zero modes (i. e., eigenstates of the Dirac operator with eigenvalues zero) that are crucial for the understanding of both the axial anomaly and spontaneous chiral symmetry breaking.

In order for instantons to be solutions of the classical equations of motions they must minimize the QCD action. First, we consider only the gauge part, that writes (cf. Eq. (2.5.6))[9]

$$\mathcal{S}_{g,\mathrm{E}} = \int \mathrm{d}^4 x \frac{1}{2g^2} \operatorname{Tr} (G_{\mu\nu} G_{\mu\nu}) \,. \tag{3.3.9}$$

One can show (cf., e.g., Ref. [58]) that this action is minimal for

$$G_{\mu\nu} = \pm \tilde{G}_{\mu\nu} \,, \tag{3.3.10}$$

i. e., the solutions of the classical Euclidean Yang-Mills theory are given by self-dual or antiself-dual fields. Furthermore, we demand a finite Euclidean action which implies that $G_{\mu\nu} \xrightarrow[|x|\to\infty]{} 0$, so that one has

$$A_\mu(x) \xrightarrow[|x|\to\infty]{} U^{-1}(x)\partial_\mu U(x) \tag{3.3.11}$$

(see Eq. (2.2.4)) for an arbitrary gauge transformation $U(x) \in \mathrm{SU}(3)_\mathrm{c}$. Such a "pure gauge field" is obtained from $A_\mu(x) = 0$ by an aforementioned gauge transformation. First solutions were found by Belavin, Polyakov, Schwartz and Tyupkin in 1975 [59].

Let us now deduce an important observation: Since the points at infinity, $|x| \to \infty$, in four-dimensional Euclidean space are three-spheres S^3 and the gauge transformations $U = \exp(-it_a\theta^a(x))$ are mappings from S^3 to SU(3) space, the topological structure of the vacuum is hence given by the *homotopy group* $\pi_3(\mathrm{SU}(3))$. It is well known that

$$\pi_3(\mathrm{SU}(N)) \simeq \mathbb{Z} \quad \text{for all } N \geq 2 \,. \tag{3.3.12}$$

From this we learn that (non-Abelian) Yang-Mills theories possess topologically degenerate vacua which can be characterized by a new quantum number (running through the domain $\mathbb{Z}$). This (topological) quantum number is the *winding number* (or *Pontryagin index*)

$$n_W = \frac{1}{16\pi^2} \int \mathrm{d}^4 x \operatorname{Tr} \left(G_{\mu\nu} \tilde{G}_{\mu\nu}\right) = \frac{1}{32\pi^2} \int \mathrm{d}^4 x\, \partial_\mu K_\mu \tag{3.3.13}$$

where we have introduced the *Chern-Simons current*

$$K_\mu := 4\varepsilon_{\mu\nu\lambda\sigma} \operatorname{Tr} \left[A_\nu\partial_\lambda A_\sigma + \frac{2}{3}A_\nu A_\lambda A_\sigma\right] \,. \tag{3.3.14}$$

Now, assuming that there are N_f light quarks present in our theory, we can relate the axial anomaly (3.3.8), on one hand, directly to the winding number (3.3.13) via

$$\int \mathrm{d}^4 x\, \partial_\mu j_5^\mu = -2N_\mathrm{f} n_W \,, \tag{3.3.15}$$

and on the other hand, to the topological current K_μ, according to

$$\partial_\mu j_5^\mu = -N_\mathrm{f} \partial_\mu K_\mu \,. \tag{3.3.16}$$

---

[9]From now on it is convenient to absorb the coupling constant g into the gauge fields, i. e., $A_\mu \to \frac{1}{g}A_\mu, G_{\mu\nu} \to \frac{1}{g}G_{\mu\nu}$.

From the first relation it follows that the axial-vector current is not conserved if $n_W \neq 0$. From the second relation, the definition of the winding number n_W and the Chern-Simons current K_μ one might argue, though, that n_W vanishes because it is given by the divergence, and hence a total derivative, of the topological current K_μ. This argument is, however, not true, because surface terms cannot be neglected after the application of Gauß's law owing to the structure of the QCD vacuum. Furthermore, K_μ is not a gauge-invariant quantity, hence having no direct physical meaning.

Instantons correspond to tunneling events between topologically different vacua. 't Hooft [33] showed that the tunneling amplitude is $T \sim e^{-S_E}$. As vacuum states $|n\rangle$ corresponding to different topological winding numbers are separated by finite-energy barriers and there is tunneling between these states, the true vacuum $|\theta\rangle$ is a superposition of these $|n\rangle$ states:

$$|\theta\rangle = \sum_n e^{-in\theta} |n\rangle . \tag{3.3.17}$$

The parameter θ labels the physically inequivalent sectors of the theory. Different θ vacua do not communicate with one another, therefore there is no *a priori* method to determine the value of θ. This θ is exactly the parameter that appears in the P- and CP-violating term (2.2.5) of the QCD Lagrangian, leading to the strong CP-problem.

Let us conclude this section by mentioning how fermions can be introduced in the instanton model (see, e. g., Ref. [58]). Fermions add to the pure gauge part of the action, $\mathcal{S}_{g,E}$ (3.3.9), the Kobayashi-Maskawa-'t Hooft determinant [32, 33] term,

$$\mathcal{S}_{int}^{(6)} = -8H \int d^4x \, [\det \mathcal{J}_+(x) + \det \mathcal{J}_-(x)] . \tag{3.3.18}$$

The coupling strength H with mass dimension -5 is defined in the same way as in the local NJL Lagrangian (3.1.2), and

$$(\mathcal{J}_\pm(x))_{ij} = \int d^4z \, \bar{\psi}_j\left(x + \frac{z}{2}\right) \frac{1}{2}(1 \mp \gamma_5) \, \mathcal{K}(z) \, \psi_i\left(x - \frac{z}{2}\right), \tag{3.3.19}$$

where $\mathcal{K}(z)$ represents the distribution of the $U(1)_A$ breaking interaction strength. For $N_f = 3$, this flavor determinant generates a genuine three-body interaction (or six-quark vertex) in which the u, d, and s quarks participate simultaneously. In his original paper [33], 't Hooft discovered the crucial property of instantons, namely the Dirac operator has a zero mode $i\slashed{D}\psi_0(x) = 0$ in the instanton field. From this observation he obtained the $U(1)_A$ breaking interaction with a simple expression for $\mathcal{K}$ as the density of zero modes. Its Fourier transform $\tilde{\mathcal{K}}(p) = \int d^4z \, e^{-ipz} \mathcal{K}(z)$ is written [58] in terms of Bessel functions and a characteristic instanton size, $d \simeq 0.35 \, \text{fm}$, as follows (compare also Fig. 3.4):

$$\tilde{\mathcal{K}}(p) = \pi p^2 d^2 \frac{d}{d\xi}\left[I_0(\xi)K_0(\xi) - I_1(\xi)K_1(\xi)\right] \qquad \text{with } \xi = \frac{|p|d}{2} . \tag{3.3.20}$$

This expression leads to a gap equation similar to that of the nonlocal two-flavor NJL model, Eq. (3.2.18). The chiral condensate in the instanton model is strongly related to fermionic zero

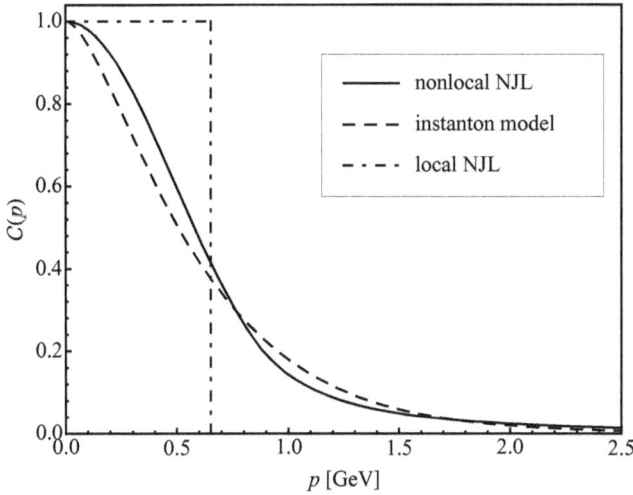

FIGURE 3.4: Comparison of the distribution $\mathcal{C}(p)$ used in the nonlocal NJL model (solid line) and the one derived from the instanton model with instanton size $d = 0.35\,\text{fm}$ (dashed line). The dot-dashed line shows the step function $\mathcal{C}(p)$ of a local NJL model with Euclidean four-momentum cutoff.

modes (Banks-Casher relation [60]).

3.3.2 Three-Flavor Nonlocal Lagrangian

We are now in the situation to write down the nonlocal equivalent to the three-flavor (local) Lagrangian (3.1.2) by combining the four- and six-fermion interaction terms (3.3.1) and (3.3.18). The four-fermion momentum distribution function $\mathcal{C}$ (cf. Eq. (3.2.39)) does in general not coincide with the instanton distribution function $\mathcal{K}$. It turns out, however, (see Fig. 3.4) that both functions are very similar, and hence we can replace $\mathcal{K}$ by $\mathcal{C}$. As outlined in detail in Appendix B, the six-fermion interaction (3.3.18) can then be expressed in terms of the currents defined in Eq. (3.3.2):

$$S_{\text{int}}^{(6)} = -\frac{H}{4} \int d^4x\, \mathcal{A}_{\alpha\beta\gamma} \left[j_\alpha^S(x) j_\beta^S(x) j_\gamma^S(x) - 3 j_\alpha^S(x) j_\beta^P(x) j_\gamma^P(x) \right], \qquad (3.3.21)$$

where the constants $\mathcal{A}_{\alpha\beta\gamma}$ are given in terms of the Gell-Mann matrices as

$$\mathcal{A}_{\alpha\beta\gamma} := \frac{1}{3!} \varepsilon_{ijk} \varepsilon_{mn\ell} \left(\lambda_\alpha \right)_{im} \left(\lambda_\beta \right)_{jn} \left(\lambda_\gamma \right)_{kl} \qquad \text{for } \alpha, \beta, \gamma \in \{0, \ldots, 8\}.$$

With the four- and six-fermion couplings in the nonlocal framework for three quark flavors, we can now write down the Euclidean nonlocal NJL action, S_{E}. We have

$$S_{\text{E}} = \int d^4x \left\{ \bar{\psi}(x) \left[-i\gamma^\mu \partial_\mu + \hat{m}_q \right] \psi(x) - \frac{G}{2} \left[j_\alpha^S(x) j_\alpha^S(x) + j_\alpha^P(x) j_\alpha^P(x) \right] + \right.$$
$$\left. -\frac{H}{4} \mathcal{A}_{\alpha\beta\gamma} \left[j_\alpha^S(x) j_\beta^S(x) j_\gamma^S(x) - 3 j_\alpha^S(x) j_\beta^P(x) j_\gamma^P(x) \right] \right\}, \qquad (3.3.22)$$

where the first term is the kinetic term and $\hat{m} = \mathrm{diag}(m_u, m_d, m_s)$ is the mass matrix with the current quark masses m_u, m_d, m_s; G and H are constants to be determined. The densities j_α^S, j_α^P are given in Eq. (3.3.2) with the nonlocality distribution $\mathcal{C}(z)$ chosen as in the two-flavor model described in Sect. 3.2.3. In the remainder of this section we demonstrate how this approach works in reproducing zero-temperature QCD, the nonperturbative vacuum and its lowest quark-antiquark excitations: the pseudoscalar meson nonet including decay constants and η-η' mixing. The bosonization procedure has already been outlined in Sect. 3.2.2. The generalization to the three-flavor action, Eq. (3.3.22), is straightforward and we simply state the result for the partition function:

$$
\begin{aligned}
\mathcal{Z} = \int \mathscr{D}\sigma_\alpha \mathscr{D}\pi_\alpha \, \det \hat{\mathscr{A}} \int \mathscr{D}S_\alpha \mathscr{D}P_\alpha \exp\left\{ \int \mathrm{d}^4x \left(\sigma_\alpha S_\alpha + \pi_\alpha P_\alpha \right) \right\} \\
\times \exp\left\{ \int \mathrm{d}^4x \left[\frac{G}{2} \left(S_\alpha S_\alpha + P_\alpha P_\alpha \right) + \frac{H}{4} \mathcal{A}_{\alpha\beta\gamma} \left(S_\alpha S_\beta S_\gamma - 3 S_\alpha P_\beta P_\gamma \right) \right] \right\}.
\end{aligned}
\tag{3.3.23}
$$

Here we have introduced 18 scalar and pseudoscalar bosonic fields σ_α and π_α ($\alpha \in \{0, \ldots, 8\}$) and, additionally, 18 auxiliary fields S_α, P_α necessary to deal with the six-fermion interactions induced by the 't Hooft term. In momentum space, the fermion determinant $\det \hat{\mathscr{A}}$ reads

$$
\mathscr{A}(p, p') := \langle p | \hat{\mathscr{A}} | p' \rangle = \left(-\not{p} + \hat{m}_q \right) (2\pi)^4 \delta(p - p') + \mathcal{C}\left(\frac{p + p'}{2} \right) \lambda_\alpha \left[\sigma_\alpha(p - p') + \mathrm{i}\gamma_5 \pi_\alpha(p - p') \right].
\tag{3.3.24}
$$

Again we conveniently write $\mathcal{C}(p) \equiv \hat{\mathcal{C}}(p)$ for the Fourier transform of the distribution $\mathcal{C}(z)$. The major differences compared with the two-flavor partition function (3.2.10) are the cubic expressions in the auxiliary fields S_α, P_α. Because of these, an analytic evaluation of the functional integral over the auxiliary fields is not possible anymore. One needs to approximate the partition function in order to get an analogous expression to the two-flavor bosonized partition function (3.2.11). This is the topic of the next subsection.

Stationary Phase Approximation

The path integration over the cubic terms in S_α and P_α fields cannot be carried out explicitly. The stationary phase approximation (SPA) is used, choosing the fields S_α, P_α so as to minimize the integrand in the bosonized partition function Eq. (3.3.23). A necessary condition imposed on the fields is therefore:

$$
\begin{aligned}
\sigma_\alpha + G S_\alpha + \frac{3H}{4} \mathcal{A}_{\alpha\beta\gamma} \left[S_\beta S_\gamma - P_\beta P_\gamma \right] = 0, \\
\pi_\alpha + G P_\alpha - \frac{3H}{2} \mathcal{A}_{\alpha\beta\gamma} S_\beta P_\gamma = 0,
\end{aligned}
\tag{3.3.25}
$$

where S_α, P_α are now to be considered as (implicit) functions of $\sigma_\alpha, \pi_\alpha$. The bosonized action can thus be written as

$$\mathcal{S}_{\text{E}}^{\text{bos}} = -\ln\det\hat{\mathscr{A}} - \int d^4x \left\{ \sigma_\alpha S_\alpha + \pi_\alpha P_\alpha + \frac{G}{2}\left[S_\alpha S_\alpha + P_\alpha P_\alpha\right] + \right.$$
$$\left. + \frac{H}{4}\mathcal{A}_{\alpha\beta\gamma}\left[S_\alpha S_\beta S_\gamma - 3S_\alpha P_\beta P_\gamma\right]\right\}. \tag{3.3.26}$$

From here on we can apply the methods developed in the two-flavor case (Sect. 3.2.2) in order to reproduce the meson spectrum within the nonlocal three-flavor NJL model.

3.3.3 Gap Equation and Meson Properties

The stationary phase approximation enables us to complete the bosonization of the nonlocal three-flavor Lagrangian in an analytical way. The price to be paid for this is, however, that one has to solve, in addition to the gap equations equivalent to Eq. (3.2.18), the SPA equations (3.3.25). By solving these equations self-consistently, we are able to reproduce the large η' mass which is the result of the anomalous breaking of the axial $U(1)_A$ symmetry. The mixing angle between the η and the η' mesons is a further outcome of this calculation.

Mean-Field Approximation, Gap Equations and Chiral Condensates

We proceed now analogously to Sect. 3.2.2. Starting from the action $\mathcal{S}_{\text{E}}^{\text{bos}}$, Eq. (3.3.26), a power series expansion is performed around the expectation values of the fields $\sigma_\alpha, \pi_\alpha$,

$$\sigma_\alpha(x) = \bar{\sigma}_\alpha + \delta\sigma_\alpha(x)\,,$$
$$\pi_\alpha(x) = \delta\pi_\alpha(x)\,. \tag{3.3.27}$$

A first constraint is imposed on the scalar fields by charge conservation, i.e., the charge matrix $\hat{Q} = \text{diag}\left(\frac{2}{3}, -\frac{1}{3}, -\frac{1}{3}\right)$ commutes with the SU(3) generators: $[\hat{Q}, \lambda_\alpha] = 0$. This is only possible for $\lambda_0, \lambda_3, \lambda_8$ which means in turn that only σ_0, σ_3 and σ_8 have to be considered (in the isospin limit which will be investigated later one has the additional constraint that σ_3 also vanishes). Given these conditions it is useful to introduce

$$\sigma = \text{diag}(\sigma_u, \sigma_d, \sigma_s) := \sigma_0\lambda_0 + \sigma_3\lambda_3 + \sigma_8\lambda_8\,, \tag{3.3.28}$$

and, analogously, $S = \text{diag}(S_u, S_d, S_s) = S_0\lambda_0 + S_3\lambda_3 + S_8\lambda_8$. Since we have $\langle\pi_\alpha\rangle = \langle P_\alpha\rangle = 0$ to leading order, the action in mean-field approximation reads

$$\frac{\mathcal{S}_{\text{E}}^{\text{MF}}}{V^{(4)}} = -2N_c \int \frac{d^4p}{(2\pi)^4}\text{Tr}\ln\left[p^2\mathbb{1}_{3\times3} + \hat{M}^2(p)\right] - \frac{1}{2}\left\{\sum_{i\in\{u,d,s\}}\left(\bar{\sigma}_i\bar{S}_i + \frac{G}{2}\bar{S}_i\bar{S}_i\right) + \frac{H}{2}\bar{S}_u\bar{S}_d\bar{S}_s\right\}, \tag{3.3.29}$$

where $\hat{M}(p) = \text{diag}\left(M_u(p), M_d(p), M_s(p)\right)$ with

$$M_i(p) = m_i + \bar{\sigma}_i\mathcal{C}(p)\,, \tag{3.3.30}$$

and $\mathbb{1}_{3\times3}$ denotes the unity matrix in flavor space and $V^{(4)}$ is the four-dimensional Euclidean volume.

The mean-field equations (gap equations) are deduced, again, by applying the principle of least action, $\frac{\delta S_{\mathrm{E}}^{\mathrm{MF}}}{\delta \sigma_i} = 0$ for $\sigma_i = \bar{\sigma}_i$ ($i \in \{u, d, s\}$). The S_i and P_i are both implicit functions of σ_i, determined through the SPA equations in mean-field approximation (compare Eq. (3.3.25)). Eventually, we obtain

$$\bar{\sigma}_i = -G\bar{S}_i - \frac{H}{4}\varepsilon_{ijk}\varepsilon_{ijk}\bar{S}_j\bar{S}_k \tag{3.3.31a}$$

$$\bar{S}_i = -8N_{\mathrm{c}} \int \frac{\mathrm{d}^4p}{(2\pi)^4} \, \mathcal{C}(p) \frac{M_i(p)}{p^2 + M_i^2(p)} \,. \tag{3.3.31b}$$

Finally, the chiral condensate $\langle\bar{q}q\rangle$ can be calculated in the same fashion as in the two-flavor case using the definition (2.4.9). This leads to

$$\langle\bar{q}q\rangle = -4N_{\mathrm{c}} \int \frac{\mathrm{d}^4p}{(2\pi)^4} \left[\frac{M_q(p)}{p^2 + M_q^2(p)} - \frac{m_q}{p^2 + m_q^2} \right] \,. \tag{3.3.32}$$

Note that $M_q(p) \to m_q$ for large p. The subtraction makes sure that no perturbative artifacts are left over in $\langle\bar{q}q\rangle$ for $m_q \neq 0$.

Second-Order Corrections and Meson Masses

The calculation of the meson masses is now more involved and more technical compared with the two-flavor case, because we are treating the (pseudoscalar) meson octet plus the η' meson which is known to have a large mass due to the anomalous breaking of the $U(1)_A$ symmetry. Calculating the masses of the pseudoscalar mesons is a major step beyond mean-field approximation. Consider second-order corrections to the mean-field action, extracted from a functional Taylor expansion,

$$\mathcal{S}_{\mathrm{E}}^{(2)} = \frac{1}{2} \int \mathrm{d}^4x \, \mathrm{d}^4y \, \frac{\delta^2 S_{\mathrm{E}}}{\delta\sigma_\alpha \delta\sigma_\beta} \delta\sigma_\alpha(x)\,\delta\sigma_\beta(y) + \frac{1}{2} \int \mathrm{d}^4x \, \mathrm{d}^4y \, \frac{\delta^2 S_{\mathrm{E}}}{\delta\pi_\alpha \delta\pi_\beta} \delta\pi_\alpha(x)\,\delta\pi_\beta(y) \,,$$

where the second derivatives, $\frac{\delta^2 S_{\mathrm{E}}}{\delta\sigma_\alpha(x)\delta\sigma_\beta(y)}$ and $\frac{\delta^2 S_{\mathrm{E}}}{\delta\pi_\alpha(x)\delta\pi_\beta(y)}$, are evaluated at the mean-field values $\sigma_\alpha(x) = \bar{\sigma}_\alpha$, etc. We now focus on pseudoscalar mesonic excitations and change the basis according to $\pi_{ij} = \frac{1}{\sqrt{2}}(\lambda_\alpha\pi_\alpha)_{ij}$. This gives a standard representation of the pseudoscalar meson octet:

$$\pi_{ij} = (\hat{\pi})_{ij} = \begin{pmatrix} \frac{\pi^0}{\sqrt{2}} + \frac{\eta_8}{\sqrt{6}} + \frac{\eta_0}{\sqrt{3}} & \pi^+ & K^+ \\ \pi^- & -\frac{\pi^0}{\sqrt{2}} + \frac{\eta_8}{\sqrt{6}} + \frac{\eta_0}{\sqrt{3}} & K^0 \\ K^- & \bar{K}^0 & -\frac{2\eta_8}{\sqrt{6}} + \frac{\eta_0}{\sqrt{3}} \end{pmatrix} \,. \tag{3.3.33}$$

Defining analogously a matrix $\hat{\sigma}$ for the scalar mesons, the fermion determinant, Eq. (3.3.24), can be written as

$$\hat{\mathscr{A}}(p, p') = \left(-\not{p} + \hat{m}_q\right)(2\pi)^4\delta^{(4)}(p - p') + \mathcal{C}\!\left(\frac{p - p'}{2}\right)\sqrt{2}\,[\hat{\sigma}(p - p') + \mathrm{i}\,\gamma_5\hat{\pi}(p - p')] \,. \tag{3.3.24'}$$

The calculation of the derivatives appearing in the Taylor expansion requires some caveats that are outlined in Appendix C. The resulting second-order contributions to the action are given by

$$S_E^{(2)} = \frac{1}{2} \int \frac{d^4 p}{(2\pi)^4} \left[G_{ij,k\ell}^+(p)\, \delta\sigma_{ij}(p)\, \delta\sigma_{k\ell}(-p) + G_{ij,k\ell}^-(p)\, \delta\pi_{ij}(p)\, \delta\pi_{k\ell}(-p) \right], \qquad (3.3.34)$$

with

$$G_{ij,k\ell}^\pm(p) = \Pi_{ij}^\pm \delta_{i\ell}\, \delta_{jk} + \left(r_{ij,k\ell}^\pm \right)^{-1}, \qquad (3.3.35)$$

where

$$\Pi_{ij}^\pm(p) = -8N_c \int \frac{d^4 q}{(2\pi)^4} \mathcal{C}^2(q) \frac{q^+ \cdot q^- \mp M_i(q^+) M_j(q^-)}{\left[q^{+2} + M_i^2(q^+) \right] \left[q^{-2} + M_j^2(q^-) \right]}, \qquad (3.3.36)$$

$q^\pm = q \pm \frac{p}{2}$, and $(r^\pm)^{-1}$ is defined as the solution of the system

$$\left[G\delta_{km}\delta_{n\ell} \pm \frac{H}{2} \varepsilon_{knt}\varepsilon_{t\ell kn} S_t \right] \left(r_{ij,k\ell}^\pm \right)^{-1} = \delta_{im}\delta_{jn}. \qquad (3.3.37)$$

The meson masses can now be determined by writing the second-order term of the action, Eq. (3.3.34), in the physical basis as

$$\begin{aligned}
S_E^{(2)}\Big|_P = \frac{1}{2} \int \frac{d^4 p}{(2\pi)^4} &\Big\{ G_\pi(p^2) \left[\pi^0(p)\, \pi^0(-p) + 2\pi^+(p)\, \pi^-(-p) \right] + \\
&+ G_K(p^2) \left[2K^0(p)\, \bar{K}^0(-p) + 2K^+(p)\, K^-(-p) \right] + \\
&+ G_{88}(p^2)\, \eta_8(p)\, \eta_8(-p) + G_{00}(p^2)\, \eta_0(p)\, \eta_0(-p) + 2G_{08}(p^2)\, \eta_0(p)\, \eta_8(-p) \Big\},
\end{aligned}$$

where the functions G_P are defined according to Eq. (3.3.35). If one considers only the isospin symmetric case, $m_u = m_d$, where $\sigma_3 = 0$, one has

$$G_\pi(p^2) = \left(G + \frac{H}{2}\bar{S}_s \right)^{-1} + \Pi_{uu}^-(p^2) \qquad (3.3.38)$$

$$G_K(p^2) = \left(G + \frac{H}{2}\bar{S}_u \right)^{-1} + \Pi_{us}^-(p^2) \qquad (3.3.39)$$

$$G_{88}(p^2) = \frac{1}{3} \left[\frac{6G - 4H\bar{S}_u - 2H\bar{S}_s}{2G^2 - GH\bar{S}_s - H^2\bar{S}_u^2} + \Pi_{uu}^-(p^2) + 2\Pi_{ss}^-(p^2) \right] \qquad (3.3.40)$$

$$G_{00}(p^2) = \frac{1}{3} \left[\frac{6G + 4H\bar{S}_u - H\bar{S}_s}{2G^2 - GH\bar{S}_s - H^2\bar{S}_u^2} + 2\Pi_{uu}^-(p^2) + \Pi_{ss}^-(p^2) \right] \qquad (3.3.41)$$

$$G_{08}(p^2) = \frac{\sqrt{2}}{3} \left[\frac{H(\bar{S}_s - \bar{S}_u)}{2G^2 - GH\bar{S}_s - H^2\bar{S}_u^2} + \Pi_{uu}^-(p^2) - \Pi_{ss}^-(p^2) \right]. \qquad (3.3.42)$$

From the construction of the action it is clear that the functions G_P correspond to the inverse pseudoscalar meson propagators. The corresponding masses are given by the poles of these propagators or, equivalently

$$G_P(-m_P^2) = 0, \quad \text{for } P \in \{\pi, K, \eta\}. \qquad (3.3.43)$$

Finally, the physical η and η' mesons can be identified after a basis change and by introducing the mixing angle $\theta = \theta(p^2)$:

$$\eta = \eta_8 \cos \theta_\eta - \eta_0 \sin \theta_\eta$$
$$\eta' = \eta_8 \sin \theta_{\eta'} + \eta_0 \cos \theta_{\eta'} \,, \tag{3.3.44}$$

where $\theta_\eta = \theta(-m_\eta^2), \theta_{\eta'} = \theta(-m_{\eta'}^2)$. Introducing the (inverse) η and η' propagators G_η and $G_{\eta'}$, respectively, instead of G_{00}, G_{88}, G_{08} we obtain for the mixing angle

$$\tan 2\theta(p^2) = \frac{2G_{08}(p^2)}{G_{00}(p^2) - G_{88}(p^2)} \tag{3.3.45}$$

and therefore:

$$G_\eta(p^2) = \frac{G_{88}(p^2) + G_{00}(p^2)}{2} - \sqrt{G_{08}^2(p^2) + \left(\frac{G_{00}(p^2) - G_{88}(p^2)}{2}\right)^2} \tag{3.3.46}$$

$$G_{\eta'}(p^2) = \frac{G_{88}(p^2) + G_{00}(p^2)}{2} + \sqrt{G_{08}^2(p^2) + \left(\frac{G_{00}(p^2) - G_{88}(p^2)}{2}\right)^2} \,. \tag{3.3.47}$$

From these formulas we conceive that there is a mass splitting, $m_{\eta'}^2 - m_\eta^2$, that is given by twice the square root appearing in Eqs. (3.3.46). In Sect. 3.3.5 we will see the quantitative impact on the η-η' mass splitting.

Renormalization Constants

Renormalized fields[10], $\tilde{\varphi}(p) = Z_\varphi^{-1/2} \varphi(p)$, are introduced so that the quadratic part of the Lagrangian can be written in the standard form

$$\mathscr{L}_{\mathrm{E}}^{(2)} = \frac{1}{2} \left(p^2 + m_\phi^2\right) \tilde{\varphi}(p) \, \tilde{\varphi}(-p) \,.$$

Masses are identified with poles of the propagators $G_P^{-1}(p^2)$. From this one arrives at an explicit expression for the renormalization constants, namely

$$Z_P^{-1} = \left.\frac{\mathrm{d}G_P(p^2)}{\mathrm{d}p^2}\right|_P , \quad \text{for } P \in \{\pi, K, \eta\} \,. \tag{3.3.48}$$

Decay Constants

The pseudoscalar meson decay constants are defined as

$$\langle 0 | J_{\mathrm{A},\alpha}^\mu(0) | \tilde{\pi}_\beta(p) \rangle = \mathrm{i} f_{\alpha\beta} \, p_\mu \quad \Longleftrightarrow \quad \langle 0 | J_{\mathrm{A},\alpha}^\mu(0) | \pi_\beta(p) \rangle = \mathrm{i} f_{\alpha\beta} Z_\beta^{1/2} \, p_\mu \,, \tag{3.3.49}$$

where $J_{\mathrm{A},\alpha}^\mu(x) = \bar{\psi}(x) \gamma^\mu \gamma_5 \frac{\lambda_\alpha}{2} \psi(x)$ denotes the axial-vector current. As described in Sect. 3.2.2 we have to gauge the nonlocal action in Eq. (3.3.22) connecting the fields by a Wilson line and introducing a set of axial gauge fields $\mathcal{A}_\mu^\alpha$ ($\alpha \in \{0 \dots, 8\}$). Following the lines of the two-flavor calculation, we find again that only the fermion determinant is affected by the gauging, becoming

---

[10]Here $\varphi(p)$ stands generically for any of the fields $\pi_\alpha(p), \dots$

then, in coordinate space,

$$
\begin{aligned}
\mathscr{A}^{\mathrm{G}}(x,y) = \left(-\mathrm{i}\,\overleftrightarrow{\partial}_y + \frac{1}{2}\gamma_5\lambda_\alpha\,\mathcal{A}^\alpha + \hat{m}_q \right)\delta(x-y) + \\
+ \,\mathcal{C}(x-y)\,\mathcal{W}\!\left(x,\frac{x+y}{2}\right)\Gamma_\alpha\,\varphi_\alpha\!\left(\frac{x+y}{2}\right)\mathcal{W}\!\left(\frac{x+y}{2},y\right).
\end{aligned}
\tag{3.3.50}
$$

Here Γ_α stands either for $\Gamma_\alpha = \lambda_\alpha$ or $\Gamma_\alpha = \mathrm{i}\gamma_5\lambda_\alpha$, and φ_α accordingly for either a scalar field, σ_α, or a pseudoscalar field, π_α.

The matrix elements (3.3.49) then follow from the gauged fermion determinant according to Eq. (3.2.27) and the result, Eq. (3.2.28), is readily generalized. The complete calculation, relegated to Appendix D, leads to:

$$
\begin{aligned}
\langle 0|J^\mu_{\mathrm{A},\alpha}(0)|\pi_\beta(p)\rangle = 2\mathrm{i}\left(\lambda^{ij}_\alpha\lambda^{ji}_\beta + \lambda^{ij}_\beta\lambda^{ji}_\alpha\right)\widetilde{\mathrm{Tr}}\left\{\mathcal{C}(q)\frac{q^+_\mu M_i(q^-)}{\left(q^{+2}+M^2_j(q^+)\right)\left(q^{-2}+M^2_i(q^-)\right)}\right\} + \\
+ 2\mathrm{i}\left(\lambda^{ij}_\alpha\lambda^{ji}_\beta + \lambda^{ij}_\beta\lambda^{ji}_\alpha\right)\widetilde{\mathrm{Tr}}\left\{\int_0^1 \mathrm{d}\alpha\, q_\mu\frac{\mathrm{d}\mathcal{C}(q)}{\mathrm{d}q^2}\frac{M_i(q^+_\alpha)}{q^{+2}_\alpha+M^2_i(q^+_\alpha)}\right\} + \\
+ 2\mathrm{i}\,\bar{\sigma}_j\left(\lambda^{ij}_\alpha\lambda^{ji}_\beta + \lambda^{ij}_\beta\lambda^{ji}_\alpha\right)\times \\
\times\,\widetilde{\mathrm{Tr}}\left\{\int_0^1\mathrm{d}\alpha\, q_\mu\frac{\mathrm{d}\mathcal{C}(q)}{\mathrm{d}q^2}\mathcal{C}\!\left(q-\frac{p}{2}\alpha\right)\frac{q^+_\alpha\cdot q^-_\alpha + M_j(q^+_\alpha)M_i(q^-_\alpha)}{\left(q^{+2}_\alpha+M^2_j(q^+_\alpha)\right)\left(q^{-2}_\alpha+M^2_i(q^-_\alpha)\right)}\right\},
\end{aligned}
\tag{3.3.51}
$$

($\widetilde{\mathrm{Tr}}$ denotes the functional trace) with

$$
\begin{aligned}
q^+_\alpha = q + \frac{p}{2}(1-\alpha), \qquad q^-_\alpha = q - \frac{p}{2}(1+\alpha) \\
q^+ = q + \frac{p}{2}, \qquad q^- = q - \frac{p}{2}.
\end{aligned}
\tag{3.3.52}
$$

The decay constants can be derived from the expression (3.3.51) and their definitions, Eq. (3.3.49), by contraction with p^μ, hence

$$
f_{\alpha\beta} = \mathrm{i}\,p_\mu\langle 0|J^\mu_{\mathrm{A},\alpha}(0)|\pi_\beta(p)\rangle\frac{Z^{-1/2}_\beta}{m^2_\beta},
\tag{3.3.53}
$$

evaluated at the corresponding mass $p^2 = -m^2_\beta$. Owing to the properties of the Gell-Mann matrices and assuming isospin symmetry (i.e., $m_u = m_d$) one has[11] $f_{\alpha\beta} = \delta_{\alpha\beta}f_\pi$ for $\alpha \in \{1,2,3\}$

---

[11]This follows from the fact, that the summands in Eq. (3.3.51) can be written as

$$
(\lambda^{ij}_\alpha\lambda^{ji}_\beta + \lambda^{ij}_\beta\lambda^{ji}_\alpha)A_{ij} = 2\,\mathrm{Re}\left(\lambda^{ij}_\alpha\lambda^{ij*}_\beta\right)A_{ij}.
\tag{3.3.54}
$$

Assuming $m_u = m_d$ one gets the stated result. Note, in particular, that for the two-flavor case, A_{ij} is independent of i, j, hence leading the anti-commutators of the Pauli matrices, as given in Eq. (3.2.28).

and $f_{\alpha\beta} = \delta_{\alpha\beta} f_K$ for $\alpha \in \{4, 5, 6, 7\}$. On the other hand, for the 0- and 8-component we obtain

$$f_{88}(p^2) = \frac{4}{3}\left[2f_{ss}(p^2) + f_{uu}(p^2)\right]$$

$$f_{00}(p^2) = \frac{4}{3}\left[2f_{uu}(p^2) + f_{ss}(p^2)\right]$$

$$f_{08}(p^2) = f_{80}(p^2) = \frac{4\sqrt{2}}{3}\left[f_{uu}(p^2) - f_{ss}(p^2)\right].$$

3.3.4 Chiral Low-Energy Theorems

In this section we are going to show that the low-energy theorems are explicitly fulfilled by the nonlocal three-flavor NJL model, as well. In order to derive the Goldberger-Treiman and Gell-Mann–Oakes–Renner relations from the nonlocal model, the meson self-energy contribution Π_{uu}^-, Eq. (3.3.36), is expanded up to first order in the current quark mass m_u and the momentum p^2:

$$\Pi_{uu}^-(p^2, m_u) = \frac{\bar{S}_{u,0}}{\bar{\sigma}_{u,0}} - \frac{2\langle \bar{u}u\rangle_0}{\bar{\sigma}_{u,0}^2}m_u + Z_{\pi,0}^{-1}p^2. \tag{3.3.55}$$

The first term on the right-hand side follows immediately from Eq. (3.3.36) by setting $m_u = 0, p^2 = 0$ and using the gap equation (3.3.31b) in the chiral limit, $m_u = 0$ (index "0"). The second term can be recovered by writing $\mathcal{C}(p) = \frac{1}{\bar{\sigma}_u}\left(M_u(p) - m_u\right)$ in Eq. (3.3.36) and using the definition of the chiral condensate, Eq. (3.3.32). The last term follows from the definition of the renormalization constant Z_π, Eq. (3.3.48).

Expanding the pion decay constant, Eq. (3.3.51), only the term in the first line of this equation contributes to order $\mathcal{O}(p^2)$. One finds

$$\lim_{p^2 \to 0} f_{uu}(p^2) = \frac{1}{4}\bar{\sigma}_{u,0}Z_{\pi,0}^{-1}. \tag{3.3.56}$$

Using Eq. (3.3.53) this implies

$$f_{\pi,0} = \bar{\sigma}_{u,0}Z_{\pi,0}^{-1}, \tag{3.3.57}$$

which is the Goldberger-Treiman relation. Finally, multiplying both sides of the expansion (3.3.55) by $G + \frac{H}{2}$, identifying the pion mass (3.3.43) on the left-hand side and the gap equation (3.3.31a) on the right-hand side, we get

$$f_{\pi,0}^2 m_\pi^2 = -m_u\langle \bar{u}u + \bar{d}d\rangle_0, \tag{3.3.58}$$

which is the Gell-Mann–Oakes–Renner relation.

We have thus demonstrated that the nonlocal three-flavor NJL model is also consistent, as expected, with fundamental low-energy theorems based on chiral symmetry. At the same time the nonlocal model reduces to the local NJL model results when $\mathcal{C}(p)$ is chosen as a theta function.

G	H	m_u	m_s
$(0.96\,\mathrm{fm})^2$	$-(0.83\,\mathrm{fm})^5$	$3.0\,\mathrm{MeV}$	$70\,\mathrm{MeV}$

TABLE 3.2: Scenario I parameter set of the $N_{\mathrm{f}} = 3$ nonlocal NJL model.

$\langle \bar{u}u \rangle = \langle \bar{d}d \rangle$	$\langle \bar{s}s \rangle$	$M_u = M_d$	M_s
$-(0.282\,\mathrm{GeV})^3$	$-(0.303\,\mathrm{GeV})^3$	$362\,\mathrm{MeV}$	$575\,\mathrm{MeV}$

m_π	m_K	m_η	$m_{\eta'}$	f_π	f_K	θ_η	$\theta_{\eta'}$
$138\,\mathrm{MeV}$	$487\,\mathrm{MeV}$	$537\,\mathrm{MeV}$	$954\,\mathrm{MeV}$	$83.4\,\mathrm{MeV}$	$99.1\,\mathrm{MeV}$	$3.3°$	$-29.1°$

TABLE 3.3: Calculated physical quantities using the scenario I parameters of Table 3.2.

3.3.5 Parameters and Results

In this section we determine the parameters of the nonlocal three-flavor NJL model in the same manner as in the two-flavor case in Sect. 3.2.4. In the present case, we fix the parameters in order to reproduce the masses and decay constants of the pseudoscalar meson nonet. Apart from the nonlocality distribution $\mathcal{C}(p)$ with its characteristic scale, the parameters to be fixed are the coupling strengths G and H and the current quark masses m_u ($= m_d$) and m_s. For the distribution $\mathcal{C}(p)$ we take over the functional form, Eq. (3.2.39), used in the two-flavor case, again (the three-flavor matching scale does not differ remarkably from the $N_{\mathrm{f}} = 2$ case: $\Gamma^{3\mathrm{f}} = 0.85\,\mathrm{GeV} \simeq (0.23\,\mathrm{fm})^{-1}$, $\Gamma^{2\mathrm{f}} = 0.83\,\mathrm{GeV} \simeq (0.24\,\mathrm{fm})^{-1}$), see Fig. 3.4.

In order to investigate the impact of the four-fermion coupling strength, we consider two scenarios with marginally different coupling strengths G (leaving $\mathcal{C}(p)$ and all remaining parameters unchanged). "Scenario I" optimizes the η' sector including η-η' mixing. "Scenario II" provides a best fit to pseudoscalar octet observables.

Choosing the parameters of scenario I as given in Table 3.2, one finds the values of the pseudoscalar masses[12], decay constants and η-η' mixing angle as shown in Table 3.3. The current quark masses are consistent with those listed in the table [19] at a renormalization scale of about $2\,\mathrm{GeV}$. The η' mass is very close to its experimental value. The same is true for the ratio of the decay constants $f_K/f_\pi = 1.19$ (compared to the experimental $(f_K/f_\pi)_{\mathrm{exp}} = 1.22$). The pion decay constant f_π, though, is approximately $10\,\%$ off its experimental value but close to its value at the chiral limit.

Furthermore, it is instructive to compare our result for the η-η' mixing angle, $\theta_{\eta'} = -29.1°$, to the empirical value. The most recent analysis [63] gives[13] $\theta = -29.0°$ and agrees perfectly with our result. Note, however, that in Ref. [63] contributions from the gluon condensate are included which our model does not explicitly account for. The left part of Fig. 3.5 shows the

---

[12]Note, that the $\bar{u}u$-threshold is lower than the η' mass. Hence, the integrals determining the η' mass might be ill-defined owing to poles in the integration region. Therefore, in fixing the η' mass, we apply the regularization method described in Appendix C.3 (following Refs. [61,62]).

[13]Note the different definitions of the η-η' mixing angle in this work and in Ref. [63]. The cited number $\theta = -29.0°$ has, however, already been translated to the definition, Eq. (3.3.44), of the mixing angle used in the present work.

G	H	m_u	m_s
$(1.04\,\text{fm})^2$	$-(0.83\,\text{fm})^5$	$3.0\,\text{MeV}$	$70\,\text{MeV}$

TABLE 3.4: Scenario II parameter set of $N_f = 3$ nonlocal NJL model.

$\langle \bar{u}u \rangle = \langle \bar{d}d \rangle$	$\langle \bar{s}s \rangle$	$M_u = M_d$	M_s
$-(0.304\,\text{GeV})^3$	$-(0.323\,\text{GeV})^3$	$468\,\text{MeV}$	$694\,\text{MeV}$

m_π	m_K	m_η	$m_{\eta'}$	f_π	f_K	θ_η	$\theta_{\eta'}$
$139\,\text{MeV}$	$495\,\text{MeV}$	$547\,\text{MeV}$	$964\,\text{MeV}$	$92.8\,\text{MeV}$	$107.5\,\text{MeV}$	$1.9°$	$-22.3°$

TABLE 3.5: Calculated physical quantities using the parameters of Table 3.4. (These values together with the parameters of Table 3.4 are referred to as "scenario II".)

momentum dependence of the resulting dynamical up-quark mass, $M_u(p)$, compared to lattice data from Ref. [53].

The parameters of scenario II (Table 3.4) differ from those of scenario I only by a four-fermion coupling constant G that is about 15 % larger. We see from the calculated quantities given in Table 3.5 that the pseudoscalar meson masses and decay constants now agree perfectly with their empirical values. On the other hand, the magnitudes of the chiral condensates and the dynamical quark masses $M_u(0)$ and $M_s(0)$ increase, making them less compatible with common phenomenology. This can easily be understood from the Gell-Mann–Oakes–Renner relation (2.4.12) recalling that the current quark and pion masses for scenario I are the same as for scenario II, while the value of the pion decay constant of scenario II is increased. The momentum dependence of the dynamical quark mass, $M(p)$, is shown on the right-hand side of Fig. 3.5. In particular, the η-η' mixing angle $\theta_{\eta'} = -22.3°$ now differs by 20 % from the deduced empirical value in Ref. [63].

As in the two-flavor case, we conclude this section with a comparison to the standard local NJL model. From Refs. [34, 64–66] one finds the gap equations

$$M_u = m_u - \tilde{G}_3 \langle \bar{u}u \rangle - \frac{\tilde{H}}{2} \langle \bar{u}u \rangle \langle \bar{s}s \rangle$$

$$M_s = m_s - \tilde{G}_3 \langle \bar{s}s \rangle - \frac{\tilde{H}}{2} \langle \bar{u}u \rangle^2 \,.$$

The equivalent coupling strengths $\tilde{G}$ and $\tilde{H}$ of the local model can be evaluated by comparison with Eq. (3.3.31a). One derives for scenario I: $\tilde{G} := G \frac{\bar{S}_u}{\langle \bar{u}u \rangle} \approx 11\,\text{GeV}^{-2}$ and $\tilde{H} := H \frac{\bar{S}_u \bar{S}_s}{\langle \bar{u}u \rangle \langle \bar{s}s \rangle} \approx 400\,\text{GeV}^{-5}$, values that lie well in the ballpark of typical local approaches [34, 35, 37, 38, 64–67].

We will comment further on the influence of the different model parameters in Sect. 5.4.3 when dealing with thermodynamics.

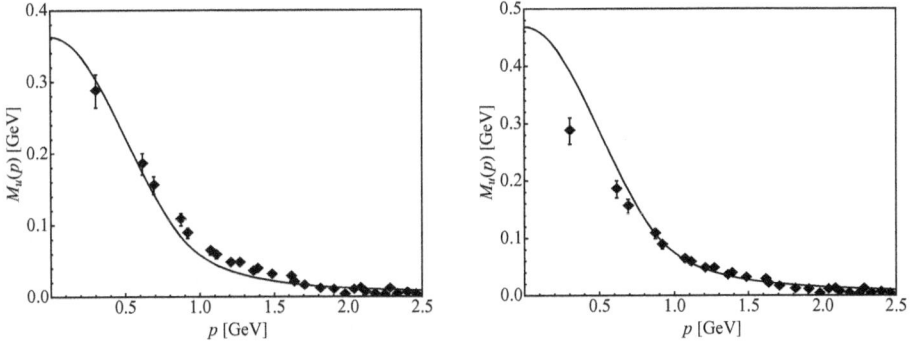

FIGURE 3.5: Momentum dependence of the dynamical (constituent) up-quark mass M_u compared with lattice data in Landau gauge extrapolated to the chiral limit (from Ref. [53] using improved staggered fermion actions). The left figure shows the result for scenario I, the right figure the result for scenario II.

3.4 Comparison of the Two- and Three-Flavor Results

We complete this chapter with a comparison of the two- and three-flavor results obtained from the corresponding nonlocal NJL models. We have calculated the dynamical (constituent) quark mass $M(p)$ (Figs. 3.2 and 3.5). It is clear, that the results are very similar. However, there are remarkable differences between two and three flavors when going to the thermodynamical treatment of the model. This is the topic of the second part of this work. Before proceeding to that, we would like to outline first merits of the model developed in the present part of the work. From its construction it is clear that the model has a strong foundation in Dyson-Schwinger calculations; the simplification made is the assumption of a separable four-fermion interaction. It has turned out, however, that effects related to differences between the full Dyson-Schwinger result and its approximation, using a separable form, can be compensated by modifying the four-point coupling strength. Thus the nonlocal NJL models incorporate, in principle, all merits of Dyson-Schwinger calculations. In particular, the integrals are convergent and do not need a cutoff regularization anymore. This is important for extensions to high temperatures and high densities, since there is no artificial length scale that limits the applicability of the model. Furthermore, by choosing an appropriate ansatz for the momentum distribution function $\mathcal{C}(p)$ we were able to reproduce the correct high-momentum asymptotics of the mass function $M(p)$. It is comforting that the width of $\mathcal{C}(p)$ is approximately the same for the two- and three-flavor model and that this width can physically be interpreted as an instanton size. Moreover, a comparison with the local NJL models is possible. It turns out, that the coupling constants G and H accord within the local and nonlocal framework, if properly translated.

We conclude that the nonlocal NJL framework presented in this chapter is consistent with constraints imposed by QCD. It provides a model that is practical and, perhaps most importantly, applicable to compute thermodynamical quantities with no principal restrictions at nonzero chemical potentials.

4 Overview of QCD Thermodynamics and QCD Phase Diagram

As a well-defined quantum field theory, QCD is, in principle, suitable to access the properties of matter under extreme conditions where the strong forces dominate. A precise knowledge and description of QCD thermodynamics is essential for the understanding of, e. g., compact stars and laboratory experiments involving relativistic heavy-ion collisions. Since the interesting experimental region of temperature T and density ϱ_B, or baryo-chemical potential μ_B, is in the range where these quantities are order Λ_{QCD}, perturbative methods do not apply. Lattice approaches which do not rely on small-parameter expansions, are the most powerful tool in studying QCD thermodynamics. Present lattice calculations at $\mu_B = 0$ lead to significant results for thermodynamic quantities in the sense that results from different lattice collaborations agree within their stated uncertainties. The description of QCD matter at $\mu_B \neq 0$ is more complicated, though. The main restriction is the so-called "sign problem" which makes it difficult to use lattice simulations at finite baryo-chemical potential. In this short chapter we describe the present status and expectations of the QCD phase diagram which best collects the phases and phase transition lines of strongly interacting matter. We highlight both recent theoretical (lattice, models) and experimental results. The detailed thermodynamical description of the model presented in the first part of this work will be redirected to the next chapter (Chapt. 5). In the present chapter we outline why model approaches are useful in order to gain, at least qualitatively, further insights into relevant topics related to the QCD phase diagram.

4.1 The Phase Diagram

Thermodynamic properties of strongly interacting matter treated as a grand-canonical ensemble are expressed in terms of a phase diagram in the plane of temperature T and baryo-chemical potential[1] μ_B. Each point on the diagram corresponds to a thermodynamic equilibrium state characterized by various thermodynamical functions such as pressure or baryon density.

In Fig. 4.1 we sketch the QCD phase diagram according to the emerging picture shared by many theorists. As already discussed in detail in Sect. 2.4.2, the vacuum of QCD is charac-

---

[1]Alternatively to the chemical potential μ_B one could also use the baryon-number density ϱ_B, which is related to μ_B and the grand-canonical potential Ω according to $\varrho_B = -\frac{\partial \Omega}{\partial \mu_B}$. The reason why μ_B is chosen is that μ_B is continuous at the phase transition lines, while ϱ_B is not. In a T-ϱ_B diagram one would obtain a transition region rather than a transition line in the T-μ_B. This can most easily be perceived if one thinks of the solid-liquid transition in water: at the melting point, temperature remains constant until the complete amount of ice has molten into (liquid) water.

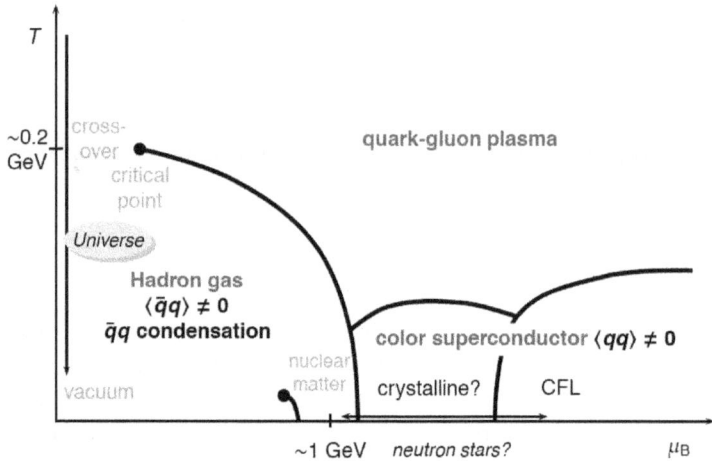

FIGURE 4.1: The contemporary view of the QCD phase diagram: both experimental and lattice data are included, on one hand, and expectations, on the other, are shown.

terized by a spontaneously broken chiral symmetry. We now investigate the different phases when temperature and density are increased. First assume that the chemical potential is small, $\mu_B \lesssim \Lambda_{QCD}$ while the temperature is increased. At sufficiently high temperature $T \gg \Lambda_{QCD}$, perturbation theory expanded around a gas of (asymptotically) free quarks and gluons should be applicable: this is the so-called quark-gluon plasma, where chiral symmetry is unbroken. One expects a transition from the spontaneously broken chiral symmetry of the vacuum state to a chirally symmetric phase at a temperature $T_c \sim \Lambda_{QCD}$.

It is more difficult to obtain some information about the order of the transition. Assuming massless quarks, the order of the phase transition (if there are massive quarks present, the phase transition turns into a crossover transition) depends on the number of flavors N_f. According to R. D. Pisarski and F. A. Wilczek [70] the transition in QCD for $N_f = 3$ massless quarks must undergo a first-order chiral restoration transition. For the $N_f = 3$ case with two massless quarks the transition can either be second- or first-order depending on the value of the strange-quark mass m_s and/or the baryo-chemical potential μ_B. The dependence of the order of (phase) transitions on the current quark masses is summarized in the so-called Columbia Plot [68], depicted in Fig. 4.2. The point on the chiral phase transition line where the transition changes its order is called tricritical point. For massless light quarks this can only be the case if the strange-quark mass is larger than m_s^{tric}, see Fig. 4.2, and with a second-order phase transition at $\mu_B = 0$. But even this statement is still under debate. Neither can it rigorously be claimed that a transition, that begins as a second-order at $\mu_B = 0$, changes to first-order, as many model and lattice calculations show.

Although not of primary interest in this work, we describe, for the sake of completeness, the low-temperature and high-density part of the QCD phase diagram. When approaching the high-baryo-chemical region, one first encounters a first-order phase transition from the hadronic

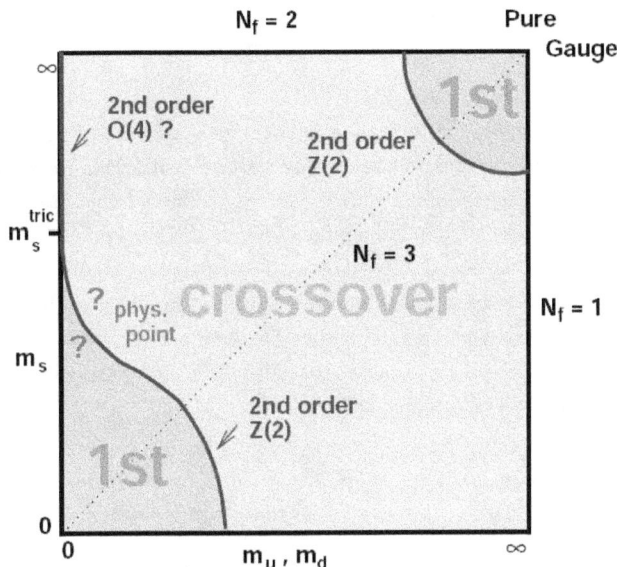

FIGURE 4.2: Illustration of (phase) transitions depending on the light- (m_u, m_d) and strange- (m_s) quark masses: Columbia plot [68]. It is still not clear, yet, where the physical point is located. The groups labeling the second-order lines correspond to the universality classes to which the second-order transitions belong: for the $N_f = 2$ chiral limit case, i. e., $m_u = m_d = 0$ one has a second-order phase transition from the chirally broken to the chirally restored phase with $O(4) \simeq SU(2) \times SU(2)$ scaling; the $(3d\text{-})Z(2)$ universality class is that of the three-dimensional Ising model. (From Ref. [69].)

gas to (liquid) nuclear matter. This is, actually, the only first-order transition found so far experimentally [71–73]. Its line ends at a critical temperature of $T_{\text{l-g}} = (17\pm3)\,\text{MeV}$ and a critical baryon density of $\varrho_{\text{l-g}} \simeq 0.95\varrho_0$, where $\varrho_0 \simeq 0.17\,\text{fm}^{-3}$ is the density of normal nuclear matter. These values are deduced from multi-fragmentation experiments [72,73]. At low temperature, nuclear matter shows spontaneous chiral symmetry breaking. When going to higher chemical potentials, recent advances [74] conjecture that nuclear matter is separated by a first-order phase transition from a color-superconducting quark matter phase. It follows from the quark-quark attraction in the color-antitriplet channel that color-flavor locked (CFL) superconductors are formed at least at asymptotically high densities.

Let us now consider the physical case where up and down quarks have a small but nonvanishing current quark mass. In this case the second-order phase transition turns into a crossover. To date, lattice calculations agree on a crossover transition at $\mu_B = 0$ [75–77], but they do not yet consistently agree on the location of the transition temperature T_c. It lies between 150 MeV [77] and 200 MeV [75], depending on the methods applied and pion masses used. Ref. [76] uses domain-wall fermions that, according to Sect. 2.4.3, do not suffer from the usual pathologies encountered for quarks on the lattice. Since such calculations require, however, an extension to a fifth dimension, the result ($T_c = 171(10)(17)\,\text{MeV}$) has still an error that can conciliate both margins for the transition temperature.

For $T \sim (1\text{--}2)T_c$ the system forms a strongly coupled quark-gluon plasma. "Strongly coupled" means that the potential energy of the system is of the same order as its kinetic energy. In the last few years, transport properties of the strongly coupled quark-gluon plasma have attracted considerable attention. The ratio of shear viscosity η to the entropy density s is of special interest. It is known to tend to infinity on either far side of the crossover (i. e., in a dilute gas and in an asymptotically free quark-gluon plasma). From empirical considerations, on the other hand [78], η/s appears to reach a minimum near the crossover. The viscosity can be indirectly determined in heavy-ion collisions by comparison of hydrodynamic calculations to experimental data. It is puzzling that why such a comparison indicates η/s nearly reaching its lower bound $\eta/s \geq 1/(4\pi)$ conjectured in holographic QCD calculations [79].

Taking into account a crossover transition for $\mu_B = 0$ (for which there is strong evidence) and a first-order transition for $T = 0$ (for which there is only weak evidence), the first-order transition line must end at a critical point. The critical (end) point of a first-order line is of second order. When approaching the end point the previously coexisting phases finally end in a single phase at that point. In QCD the coexisting phases along the first-order line are a hadron gas and the quark-gluon plasma. The distinction of the two phases is only quantitative: chiral symmetry, on one hand, is explicitly broken by quark masses, therefore, the two phases cannot be distinguished by realizations of any global symmetry; deconfinement, on the other hand, does not provide a strict distinction from the confined phase, either. With quarks, even in vacuum, the confining potential cannot rise infinitely, because a quark-antiquark pair inserted into the color flux tube breaks it. The energy required to separate two color charges from each other is finite.

It has been conjectured recently that the critical point in the QCD phase diagram should rather be a triple point at which hadronic matter, the quark-gluon plasma and a new state of matter referred to as "quarkyonic matter" coexist [80, 81]. Quarkyonic matter is (approximately) confined and it is characterized by a large energy density due to quarks deep in the Fermi sea. Whether chiral symmetry is broken or restored cannot be answered unequivocally, because there is always the possibility for chiral symmetry breaking even at high densities from pairing effects near the Fermi surface. As an experimental indication for the possible existence of an additional state of matter one considers the chemical freeze-out temperatures at large μ_B. From experiments at the Relativistic Heavy Ion Collider (RHIC) one deduces that chemical freeze-out takes place close to the phase boundary (at low μ_B), driven by the rapid density change across the phase transition. At high μ_B, the confinement-deconfinement transition is far apart from the freeze-out line, but this still takes place within a small interval in temperature and chemical potential for all hadrons. From this it is argued that one needs an additional phase. Its transition line to the hadronic state is believed to provide the rapid change in density, necessary for the sharp chemical freeze-out region.

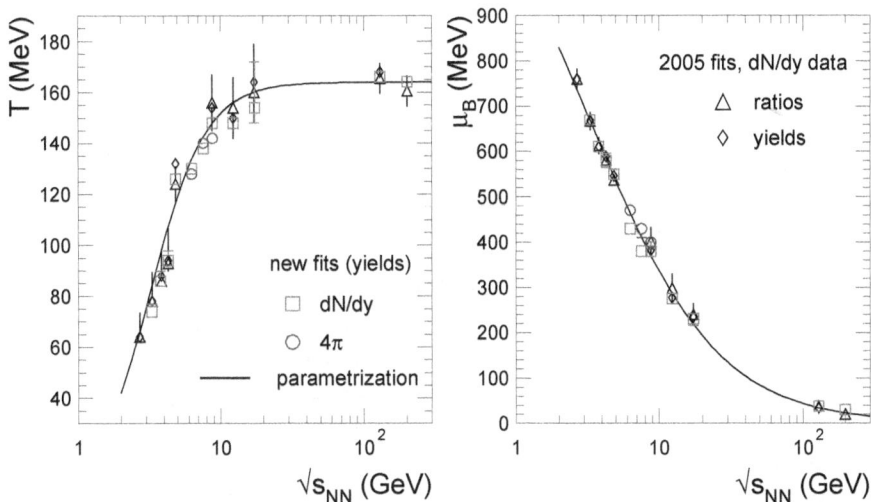

FIGURE 4.3: Temperature and baryo-chemical potential of Statistical Model fits to hadro-chemical abundances as a function of center of mass energy per nucleon pair for collisions of heavy nuclei from Refs. [82, 83].

4.2 Experimental Status

While the existence of a critical point in the QCD phase diagram[2] (apart from the one associated with the nuclear liquid-gas transition) is not yet established, we mention in this short section briefly how one could detect signatures for a critical point. It is known empirically that with increasing collision energy $\sqrt{s}$ the generated fireball freezes out at decreasing values of the chemical potential (see Fig. 4.3). This is easily understood by the fact that the amount of generated entropy grows with $\sqrt{s}$ while the net baryon number is limited by that in the initial nuclei. Simulating the time evolution of a fireball by ideal hydrodynamics, the trajectories follow lines of constant baryon number per entropy density because of baryon number and entropy conservation. What is more, these lines point towards the critical end point, a phenomenon known as *focusing*: the (chemical) freeze-out points[3] tend to cluster near the critical point. Hence, it suffices to hit a neighborhood around the critical point in heavy-ion collisions. Experimentally, the location of the freeze-out point is obtained by measuring the ratios of particle yields (e. g., baryons or antibaryons to pions) and fitting to statistical models. Fig. 4.7 shows some freezeout points for orientation.

One searches for signatures of the critical point by considering the non-monotonous dependence on $\sqrt{s}$ (and hence, on μ_B) of event-by-event fluctuation observables. This idea relies on the fact

---

[2]Here we mean, as before as well, the end point of the hadronic gas to quark-gluon plasma state and not the critical point at the end of the liquid-gas transition (between the hadron gas and nuclear matter) at low temperature.

[3]These are the points in T-μ_B plane below which the hadron content does not change anymore.

that susceptibilities diverge at the critical point and that the magnitude of the fluctuations is proportional to the corresponding susceptibilities. As a starting point, one considers the quark number susceptibilities

$$\langle \Delta n_{\vec{p}}^i \Delta n_{\vec{k}}^j \rangle = \langle n_{\vec{p}}^i n_{\vec{k}}^j \rangle - \langle n_{\vec{p}}^i \rangle \langle n_{\vec{k}}^j \rangle \,, \qquad (4.2.1)$$

where $\Delta n_{\vec{p}}^i := n_{\vec{p}}^i - \langle n_{\vec{p}}^i \rangle$ denotes the event-by-event fluctuation of the number of particles of the type i in the momentum bin centered around $\vec{p}$. Although this quantity can be measured directly, one often uses so-called cumulative measures which sum over all momenta. The charge fluctuations are then given, e. g., by $\Delta Q = \sum_{\vec{p},i} q^i \Delta n_{\vec{p}}^i$, where q^i is the charge of the particle i. In our recent work [84] we calculated quark number susceptibilities for the local Nambu–Jona-Lasinio models. Such quantities are useful also for comparison to experiment, once data will be available. In addition to charge fluctuation one often considers quark- or baryon-number fluctuations or transverse-momentum fluctuations. The characteristic feature of all such signatures is the non-monotonic dependence on the value of an experimentally controlled parameter, such as the center of mass energy $\sqrt{s}$, as the critical region is entered. Near criticality, the crucial quantity is the value of the correlation length ξ. In experiments, the divergence of ξ is, however, limited by two effects: first, the finite system size of order 10 fm for heavy-ion collisions, and secondly, the finite evolution time τ. It turns out [85, 86], that the second effect is more important. The time during which the correlation length reaches equilibrium diverges as $\tau \sim \xi^z$, which defines the *dynamic* scaling exponent z. This definition allows one now to conclude the discussion of Sect. 4.1 on the universality arguments entering the QCD phase diagram. As mentioned there, it has been well established that the *static* universality class of the QCD critical point is that of the three-dimensional Ising model. It was proven, however, only recently [87], that the dynamic universality class of the critical point is the same as that for the liquid-gas transition (the universality class of model H in Hohenberg and Halperin's classification [88]). Hence we see, that equal *static* universality classes (three-dimensional Ising model and liquid-gas transition) *do not* necessarily imply the same dynamical universality classes (3d Ising model is in the model A class, the liquid-gas transition in the H-model class according the the classification of Ref. [88]).

Apart from these general considerations, we do not comment on possible experimental signatures further, because a detailed description of present experiments is beyond the scope of this work. On the other hand, a reasonable approach towards the critical region can probably be made only starting with the low-energy run at RHIC (starting from 2010), at the CBM experiment at FAIR (~ 2020) and with the ALICE experiment at LHC that extends the data base to the highest possible energies.

4.3 Lattice-QCD Calculations at Finite Temperature and Density

Finite-temperature and -density lattice calculations use Monte Carlo methods in order to compute the thermodynamical partition function, to be defined in Eq. (5.1.1), in its path-integral representation, Eq. (5.1.3). This allows lattice collaborations to determine the equation of state of QCD as a function of T for vanishing baryo-chemical potential, $\mu_B = 0$. One finds a crossover with a transition temperature T_c in the range between 150 MeV [77] and 200 MeV [75], as already discussed in the previous section. Although there is still an uncertainty in the location of T_c, there is no doubt about *the existence* of a crossover transition at zero density from lattice simulations.

4.3.1 Lattice Methodology at Finite Density

At finite μ_B, lattice calculations become more involved due to the notorious sign problem: calculating the partition function using Monte Carlo methods relies on the fact that the exponent of the Euclidean action $\mathcal{S}_E$ is a positive-definite function of its variables. The so-called *importance sampling* allows one then to limit the calculation to a tractably small set of random field configurations needed to achieve reasonable accuracy. Importance sampling relies on the fact that only those of the $e^{\text{const} \cdot V}$ configurations are important for which $e^{-\mathcal{S}_E}$ is sizeable. In finite-density QCD the situation changes dramatically: from Eq. (5.1.1) we see that a nonzero chemical potential modifies the zeroth component of the four-momentum vector in Minkowski space, which, in the Euclidean- or imaginary-time formalism translates into a shift of p_4 into the imaginary Euclidean time direction by $i\mu_B$. Therefore, the action $\mathcal{S}_E$ is in general complex. This is obviously a major impediment for importance sampling because there is no ordering according to which complex numbers could be compared with one another. Several methods to circumvent this issue have been used, but none of them can be expected to converge to the correct result when increasing the lattice volume. This can be best shown by considering the most straightforward method for finite chemical potential: reweighting. In this case, one uses the $\mu_B = 0$ sample even at $\mu_B \neq 0$ and corrects for the incorrect probability measure by multiplying the contribution of each configuration by $e^{\mathcal{S}_E|_{\mu_B=0} - \mathcal{S}_E}$. This is exact in the limit when all possible configurations would be considered. But when considering a finite volume V and a finite number of samples, one relies on the hope that configurations important for $\mu_B \neq 0$ are the same as for $\mu_B = 0$. The probability for this to happen is exponentially small, $e^{-\text{const} \cdot V}$, and the reweighting factor, correcting for this, is exponentially large ("overlap problem"). Hence, the obtained result suffers from large fluctuations that wash out the significance of the result.

Irrespective of these caveats, the first lattice prediction for the location of the critical point was reported by Fodor and Katz [89, 90]. From Fig. 4.7 one sees that the critical point found in Refs. [89, 90] is at a small value of μ_B, thus there is hope that the volume V might not be too large in order to achieve a reasonable accuracy.

Another method which allows for dealing with the imaginary fermion determinant was proposed by P. de Forcrand and O. Philipsen [91, 92] and uses the continuation to an imaginary chemical potential μ_B. In this case the extra term $i\mu_B$ introduced to the partition function when dealing with finite densities is again real and consequently the partition function itself is a real quantity. It is indeed not devious to consider an imaginary μ_B as may be conceived from the following reasoning: we mentioned earlier that in the case of three vanishing current quark masses the system undergoes a first-order phase transition at $\mu_B = 0$. By continuity arguments, this must be true for a finite region around zero current quark masses (cf. the Columbia plot

Fig. 4.2). Hence, when decreasing the current quark masses one expects the critical point to be pulled towards the $\mu_B = 0$ axis until it disappears off the phase diagram. Equivalently, this means that the "critical point" then appears at imaginary μ_B not causing any problems in the calculation of the partition function on the lattice anymore. De Forcrand and Philipsen have studied the μ_B^2 dependence of the quark masses, in particular of the strange-quark mass $m_s = m_s(\mu_B^2)$ for $\mu_B^2 < 0$, and have then analytically continued to real μ_B which has enabled them to estimate the position of the critical point in the T-μ_B plane. In Ref. [93] de Forcrand and Philipsen determine with the method of imaginary chemical potential the curvature $\frac{\mathrm{d}^2 m_c}{\mathrm{d}\mu_B^2}$ of the *critical surface* at $\mu_B = 0$. Here m_c denotes the critical mass that determines the critical point $(T(m_c), \mu_B^2(m_c))$. The critical surface can be thought of as the surface separating the first-order and crossover transitions when extending the Columbia plot to nonzero chemical potentials. In Ref. [93] it is found that this surface has negative curvature, meaning that no critical point would be present in the QCD phase diagram. There are no general arguments that forbid the critical surface having negative curvature (it turns out, that this question is related to the strength of the axial anomaly [94]), nor do there exist any constraints that hinder a change in the sign of the curvature, i. e., it could well be that the critical curvature bends into the opposite direction at higher chemical potentials.

Another method to circumvent the sign problem is a Taylor expansion of the thermodynamical potential Ω in μ_B. This permits calculating the μ_B dependence of, e. g., the baryon-number susceptibility $\chi_B \sim \frac{\mathrm{d}^2\Omega}{\mathrm{d}\mu_B^2}$. This quantity is of particular interest because it serves as a signature of the critical point in experiments: like in ferromagnets, the susceptibility diverges at the critical point[4]. Finally, such Taylor expansions possess a finite radius of convergence that is determined by the nearest singularity in the complex μ_B plane. Assuming that the critical point is the nearest singularity to μ_B one can use the radius of convergence in order to find an estimate for the critical point.

4.3.2 Recent Lattice-QCD Developments

QCD thermodynamics on the lattice has recently been simulated essentially by two groups: first, the Budapest-Marseille-Wuppertal collaboration [77, 95], and second, the "hotQCD" collaboration [75, 96] (which is composed by members from RIKEN, Brookhaven National Laboratory, Columbia and Bielefeld — RBC-Bielefeld collaboration, and part of the MILC collaboration). Before discussing some of the details of the simulations we first describe the general formulation that underlies all lattice calculations at finite temperature.

In a lattice calculation, temperature and volume are given in terms of the temporal (N_τ) and spatial (N_σ) extent of the four-dimensional discretized Euclidean space. The lattice spacing a is controlled by the lattice gauge coupling $\beta := 2N_c/g^2$ ($\beta = 6/g^2$ for $N_c = 3$) and is related to temperature T and lattice volume V through

$$T = \frac{1}{N_\tau a(\beta)}, \qquad V = (N_\sigma a(\beta))^3. \qquad (4.3.1)$$

---

[4]Remember, that the critical point of a first-order phase transition is of second order.

The g denotes the coupling strength. Note that the lattice coupling β must not be confused with the QCD beta function (2.3.7) encountered in the discussion of the renormalization group equations.[5] All observables calculated on the lattice are functions of the coupling β. This has important consequences for lattice simulations at finite temperature: the coupling strength β is related, on one hand, to the QCD Lagrangian or action at zero temperature and, on the other hand, through Eq. (4.3.1) to the temperature. This means that simulating a system at a particular temperature T on the lattice requires the determination of a certain lattice spacing $a(\beta)$ as a function of the coupling β (where a constant N_τ is assumed).

The QCD (grand-canonical) partition function on a lattice of size $N_\sigma^3 N_\tau$ is

$$\mathcal{Z}_{\mathrm{LCP}}(\beta, N_\sigma, N_\tau) = \int \prod_{x,\hat{\mu}} \mathrm{d}U(x; \hat{\mu})\, \mathrm{e}^{-\mathcal{S}_{\mathrm{E}}(U)}, \tag{4.3.2}$$

where $U(x; \hat{\mu}) \in \mathrm{SU}(3)$ denotes the gauge link variables connecting lattice sites x and $x + \hat{\mu}$ encountered in Sect. 2.4.3. The Euclidean action, $\mathcal{S}_{\mathrm{E}}$, decomposed into its gauge and fermionic part is given, respectively, by

$$\mathcal{S}_{\mathrm{E}}(U) = \beta \mathcal{S}_{\mathrm{G}}(U, \beta) - \mathcal{S}_{\mathrm{F}}(U, \beta). \tag{4.3.3}$$

The subscript LCP in Eq. (4.3.2) indicates that the partition function is defined on a *line of constant physics*. The LCP is defined at $T = 0$ as a line in the space of light- and bare quark masses, $m_l := m_u = m_d$ and m_s, respectively, parametrized by the coupling β. Each point on this line corresponds to identical physical conditions at different values of the lattice spacing $a(\beta) \propto \exp\left(-\frac{\beta}{4N_c b_0}\right)$, $b_0 = \frac{9}{16\pi^2}$, which is tuned towards the continuum limit (i.e., for $a \to 0$ or $\beta \to \infty$) by increasing the gauge coupling β. Since each temperature is related to a particular value of β according to Eq. (4.3.1), the current quark masses have to be adjusted accordingly. The two publications considered here use different physical observables that are fixed in their zero-temperature runs for the determination of the LCP. We will discuss them later in this subsection. Finally, the LCP introduces a physical scale into the lattice calculations that permits to define the temperature scale for the thermodynamics calculations.

From the partition function (4.3.2) one can easily calculate the grand-canonical potential $\Omega(T, V)$, normalized such that it vanishes at zero temperature,

$$\Omega(T, V) = -T \ln \mathcal{Z}(T, V) - \Omega_0 \tag{4.3.4}$$

with $\Omega_0 := \lim_{T \to 0} \left[-T \ln \mathcal{Z}(T, V)\right]$. This prescription is equivalent to requiring a zero pressure or energy density at vanishing temperature.

In this subsection the primary interest is in the recent discussion concerning the transition temperature from the chirally broken to the chirally restored phase at vanishing baryo-chemical potential. The appropriate order parameter for chiral symmetry breaking is the chiral condensate

---

[5]One finds, though, for the weak-coupling limit (i.e., $\beta \to \infty$): $T\frac{\mathrm{d}\beta}{\mathrm{d}T} = 12b_0 + 72b_1/\beta + \mathcal{O}(\beta^{-2})$ with $b_0 = 9/(16\pi^2)$ and $b_1 = 1/(4\pi^4)$. Thus, $T\frac{\mathrm{d}\beta}{\mathrm{d}T}$ approaches the universal form of the two-loop β function of three-flavor QCD.

$\langle \bar{q}q \rangle$, $q \in \{u, d, s\}$. It is nonzero at low temperature and vanishes above a critical temperature T_c. Chiral symmetry is broken spontaneously for $T < T_c$. The condensate $\langle \bar{q}q \rangle$ is, however, an exact order parameter only in the chiral limit; in the presence of nonvanishing (light) current quark masses, the phase transition turns into a smooth crossover. Nevertheless, the chiral condensates, given as the derivatives of the thermodynamical potential with respect to quark masses,

$$\langle \bar{q}q \rangle = \frac{T}{V} \frac{\partial \ln \mathcal{Z}}{\partial m_q}, \qquad q \in \{u, d, s\}, \qquad (4.3.5)$$

are used to trace the transition.

The chiral condensates need to be renormalized. At zero quark mass a multiplicative renormalization is sufficient. At nonzero values of the quark mass, an additional renormalization is necessary to eliminate singularities that are proportional to m_q/a^2. An appropriate renormalized order parameter for chiral symmetry breaking can be defined as

$$\Delta_{l,s}(T) := \frac{\langle \bar{u}u \rangle_T - \frac{m_u}{m_s} \langle \bar{s}s \rangle_T}{\langle \bar{u}u \rangle_0 - \frac{m_u}{m_s} \langle \bar{s}s \rangle_0}, \qquad (4.3.6)$$

where the subscripts on the angled brackets, $\langle \cdot \rangle_T, \langle \cdot \rangle_0$, indicate that the quantities are calculated at temperature T or zero temperature, respectively. $\Delta_{l,s}$ is unity at low temperatures and vanishes at T_c for $m_u = m_d = 0$.

The derivative of the chiral condensate with respect to the quark mass defines the chiral susceptibilities

$$\chi_{m,q} = \frac{T}{V} \frac{\partial^2 \ln \mathcal{Z}}{\partial m_q^2}, \qquad q \in \{u, d, s\}. \qquad (4.3.7)$$

The divergence of $\chi_{m,q}$ at T_c *in the chiral limit* is an unambiguous signal of the chiral phase transition. For finite quark masses, $\chi_{m,q}$ is expected to show a pronounced peak structure enabling one to determine a crossover transition temperature T_c.

Apart from the chiral transition, the bulk thermodynamic observables are determined by the transition between the confined and deconfined phase. The sudden liberation of partonic degrees of freedom in QCD manifests itself in a rapid change of thermodynamical quantities. It turns out [97, 98] that a suitable order parameter for this confinement-deconfinement transition is the expectation value $\langle \Phi \rangle$ of the Polyakov loop $L(\vec{x})$,

$$\langle \Phi \rangle := \frac{1}{N_c} \left\langle \frac{1}{N_\sigma^3} \sum_{\vec{x}} L(\vec{x}) \right\rangle \qquad \text{with} \qquad L(\vec{x}) := \text{Tr} \prod_{x_4=1}^{N_\tau} U\left((x_4, \vec{x}); \hat{n}_4\right), \qquad (4.3.8)$$

where $\hat{n}_4$ denotes the unit vector in four-direction on the lattice. In Sect. 5.2 it will be shown that the Polyakov loop is related to the free energy, $\mathscr{F}_\infty$, of a quark-antiquark pair with infinite spatial separation (cf. Refs. [97, 98]):

$$|\langle \Phi(T) \rangle|^2 = e^{-\frac{1}{T} \mathscr{F}_\infty(T)}. \qquad (4.3.9)$$

From this it is clear, that $\langle \Phi \rangle$ vanishes for the confined phase, where $\mathscr{F}_\infty \to \infty$, while it is nonzero for the confined phase characterized by a finite free energy. Some remarks are in order: first, the Polyakov loop is an exact order parameter for deconfinement only for the pure gauge theory, i.e., for all quark masses taken to infinity (see Sect. 5.2). At finite quark masses it is nonzero for all values of the temperature but changes rapidly at the transition such it can still be used for an approximative determination of T_c. Second, the Polyakov-loop operator is not present in the QCD action but can be added to it as an external source. Its expectation value is then given by the derivative of the modified thermodynamic potential with respect to the corresponding coupling, evaluated at zero coupling. Third, the Polyakov loop needs to be renormalized in order to eliminate self-energy contributions to the static-quark free energy. The detailed renormalization prescription depends on the lattice action used. We will not comment on this further here.

We proceed now to compare the results of the Budapest-Marseille-Wuppertal collaboration [77, 95], and of the "hotQCD" collaboration [75, 96], both obtained for the (2+1)-flavor case. The left picture of Fig. 4.4 and Table 4.1 demonstrate discrepancies in the temperature dependence of the chiral order parameter $\Delta_{l,s}$. We discuss briefly the methods applied in the two collaborations and comment at the end about possible reasons for the different values of the transition temperature T_c given by these two groups.

The Budapest-Marseille-Wuppertal collaboration has investigated, in a series of the three publications [77, 95, 99], the impact of the size of the current quark masses and the lattice spacing. Refs. [77, 99] are a major extension of a previous study in Ref. [95]. The simulations in all three references were performed using a Symanzik-improved gauge and a stout-link-improved staggered fermionic lattice action. The more recent publications [77, 99] are of particular interest because they use, for the first time, physical quark masses. This means that in [77, 99] the LCP, and hence $m_{u,d} = m_{u,d}(\beta), m_s = m_s(\beta)$, were determined such that the pion and kaon masses and the kaon decay constant assume their physical values. In contrast, in Ref. [95] the aforementioned quantities were calculated at higher current quark masses rather than the physical ones. Different fit formulas were then used in order to extrapolate the ratios m_K/m_π and m_K/f_K to their physical values. (It turns out, that the direct determination and the extrapolation agree within 2 %.) The lattice scale, that converts lattice units into physical units, was also fixed by the pion and kaon masses, and by the kaon decay constant (a change of the experimental value of the kaon decay constant, f_K in 2008 resulted in a reduction of T_c by about 6 MeV in Refs.'s [77, 95] predictions in Table 4.1). The finite temperature simulations in Ref. [77] were performed on lattices up to $N_\tau = 12$ and $N_\sigma = 36$ (extending the $N_\tau = 4, 6, 8, 10$ calculations in Ref. [95]) while Ref. [99] adds to them the results of an $N_\tau = 16$ calculation. The temperature dependence of the quantities given in Table 4.1 was determined. As a generic feature of any crossover, the transition temperatures obtained from different quantities are different. Fig. 4.4 shows the temperature dependence of the chiral order parameter $\Delta_{l,s}$ (left picture, from Ref. [99]) and of the (renormalized) Polyakov loop[6] (right picture, from Ref. [99]). After taking the continuum

---

[6]In order to compare the results of Ref. [99] to those of the "hotQCD" collaboration a new definition (compared to Refs. [77, 95]) for the Polyakov loop was applied. Therefore, a direct comparison with Refs. [77, 95] is not possible, and thus the slight difference in the transition temperatures found in Refs. [77] and [99].

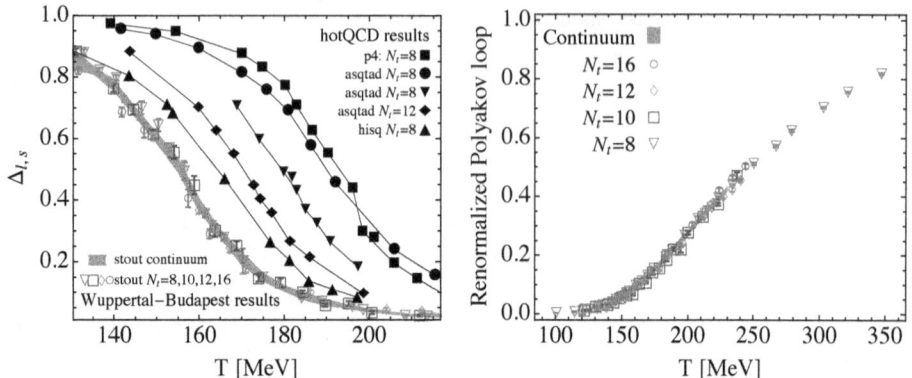

FIGURE 4.4: Left: Renormalized chiral condensate $\Delta_{l,s}$ as a function of the temperature. Colored open symbols are the (stout) results on $N_\tau = 8, 10, 12$ and 16 lattices from Ref. [99]; the gray band is their continuum result. For comparison, results of the "hotQCD" collaboration with three different fermion actions are also shown (compare Fig. 4.5): the $N_\tau = 8$ p4 (full squares) and asqtad (full circles) results from Ref. [96] were calculated for $m_\pi = 220$ MeV, the other asqtad results (full up triangle and full diamond) as well as the HISQ result (full down triangle) correspond to a quasi-physical $m_\pi = 160$ MeV. Right: Temperature dependence of the renormalized Polyakov loop for $N_\tau = 8, 10, 12$ and 16 from Ref. [99]. The gray band is the continuum extrapolated result.

limit, the temperatures range from 146 MeV to 170 MeV.

Consider next the lattice simulations of the "hotQCD" collaboration. Ref. [96] provides an extension of the the results from Ref. [75] to $N_\tau = 8$ over a large temperature range $T \in [100 \text{ MeV}, 500 \text{ MeV}]$ using a Symanzik-improved gauge and two different improved staggered-fermion actions, the asqtad and p4 actions. Refs. [75, 96] define the line of constant physics by demanding that the ratio of the strange-pseudoscalar and the kaon mass, $m_{\bar{s}s}/m_K$, stays constant[7] and $m_{\bar{s}s}$ expressed in terms of the Sommer scale[8] r_0 stays constant. Using $m_{\bar{s}s}/m_K = 1.33$, $m_l/m_s = 0.1$ and $r_0 = 0.469$ fm one obtains $m_\pi \simeq 220$ MeV, $m_K \simeq 503$ MeV and $m_{\bar{s}s} \simeq 669$ MeV. The temperature dependence of the chiral order parameter $\Delta_{l,s}$ and the (renormalized) Polyakov loop L_{ren} from Ref. [96] are shown in Fig. 4.5. While Ref. [75] cites a transition temperature $T_c = 196(3)$ (for $N_\tau = 6$) a more recent publication [96] relaxes this number to a broad transition region $180 \text{ MeV} \leq T_c \leq 200 \text{ MeV}$. The results obtained using the asqtad and p4 actions for different values of the lattice cutoff, $aT = 1/N_\tau = 1/6$ and $1/8$, are subject to discretization effects: in the vicinity of the transition region they amount to 10% and at most a few percent for temperatures larger than 300 MeV. For this reason, the calculations of the

---

[7]This is, indeed, a reasonable assumption if one assumes the ratio m_l/m_s staying constant. Using Gell-Mann–Oakes–Renner-like relations for $m_{\bar{s}s}$ and m_K, i.e., $m_{\bar{s}s}^2 \sim m_s$ and $m_K^2 \sim m_l + m_s$, respectively, one easily sees that the ratio $m_{\bar{s}s}/m_K$ is a constant. Refs. [75, 96] use $m_l/m_s = 0.1$.

[8]The Sommer scale is defined as the distance at which the slope of the zero-temperature, static quark potential $V_{\bar{q}q}(r)$ assumes a certain value:

$$\left(r^2 \frac{dV_{\bar{q}q}(r)}{dr} \right)_{r=r_0} = 1.65 \,.$$

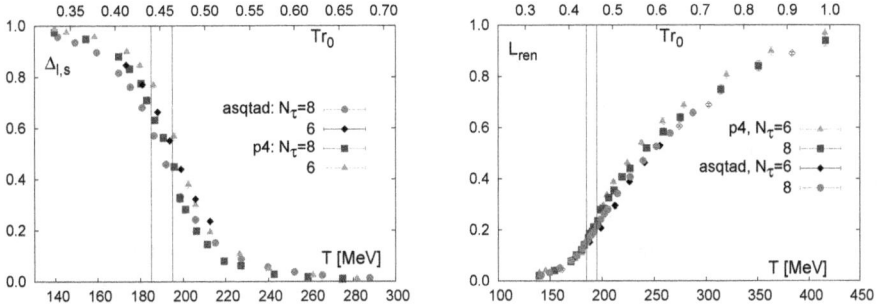

FIGURE 4.5: Temperature dependence of the chiral order parameter $\Delta_{l,s}$ (left) and the renormalized Polyakov loop L_{ren} (right) obtained with the asqtad and p4 actions from simulations on lattices with temporal extent $N_\tau = 6$ and 8 (from Ref. [96]).

"hotQCD" collaboration have been advanced with respect to more improved staggered-quark action and lower pion masses. The p4-action calculation of Ref. [100] uses a physical strange-quark mass and performs the lattice calculations with $m_l/m_s = 0.05$, leading to a pion mass of $m_\pi = 154\,\mathrm{MeV}$ and a kaon mass $m_K = 486\,\mathrm{MeV}$. The result for the chiral condensate $\Delta_{l,s}$ is shown in the left picture of Fig. 4.4.[9] Ref. [101] uses the Highly Improved Staggered Quark (HISQ) action for a light-to-heavy-current-quark ratio $m_l/m_s = 0.05$ (giving $m_\pi = 158$–$160\,\mathrm{MeV}$ and $m_K = 496$–$504\,\mathrm{MeV}$). The HISQ action reduces the taste symmetry breaking due to the removal of tree level $\mathcal{O}(a^2)$ artifacts and, hence, decreases the splitting between different pion tastes by a factor of about three compared to the asqtad action. The reduced lattice artifacts lead to a more realistic hadron spectrum. The results obtained for the chiral order parameter $\Delta_{l,s}$ are shown in Fig. 4.4. They clearly show a strong decrease of the transition temperature to a region around $T_c = 170\,\mathrm{MeV}$. Furthermore, Fig. 4.4 shows that the asqtad results on $N_\tau = 12$ lattices agree well with the ones obtained with the HISQ action on $N_\tau = 8$. On the other hand, the HISQ results disagree with the results obtained with the p4 and asqtad actions on $N_\tau = 8$ lattices.

In summary, the Wuppertal-Budapest and "hotQCD" results for the chiral transition temperature disagree by approximately 20–35 MeV. From Fig. 4.4 and Table 4.1 it is clear that all characteristic temperatures for the "hotQCD" collaboration are higher. Ref. [99] compares the results of both the Wuppertal-Budapest collaboration and the "hotQCD" collaboration to a hadron-resonance-gas (HRG) model. This model has its origin in a theorem by Dashen, Ma and Bernstein which allows one to calculate the microcanonical partition function of an interacting system, in the thermodynamic limit, i.e., $V \to \infty$, assuming that it is a gas of noninteracting free hadrons and resonances. The HRG version of Ref. [102] permits to describe the pion-mass and lattice-spacing dependence of the hadron masses. Combining this with results from chiral perturbation theory, it turns out [99] that the Wuppertal-Budapest results are in complete agreement with the HRG model (below the transition region) using the physical hadron spectrum.

---

[9]In order to be precise, Fig. 4.4 shows the p4 result for $m_\pi = 220\,\mathrm{MeV}$. The result for $m_\pi = 154\,\mathrm{MeV}$, as given in [100] is shifted by about 5 MeV towards lower temperatures.

The results of the "hotQCD" collaboration can, however, be reproduced within the HRG model if one uses a "distorted" spectrum which takes into account the larger quark masses, as well as the larger lattice spacing and pseudoscalar meson splittings. This analysis therefore provides a convincing explanation of the observed shift in transition temperatures between the two collaborations. Another issue is related to the determination of the physical scale. As mentioned earlier, the Wuppertal-Budapest collaboration uses, apart from the pion and kaon masses, the kaon decay constant, f_K, while the "hotQCD" collaboration uses the Sommer scale r_0 for the translation from lattice to physical units. In Ref. [101] it is pointed out that the stout results at $N_\tau = 12$ differ by about 10% depending on whether f_K or r_0 has been used. On the other hand, Refs. [77, 95] argue that only continuum extrapolated results are physical, and that using f_K and r_0 scale settings give the same continuum result. This emphasizes the important role of taking the proper continuum limit.

Regardless of the discussion of the transition temperature in the framework of staggered-fermion actions we mention here that the staggered formalism and all other large scale thermodynamics studies may suffer from theoretical problems. To date it is not proven that the staggered formalism with $2 + 1$ flavors really describes QCD in the continuum limit. Staggered fermions suffer from the disadvantage that they do not preserve the full $SU(2) \times SU(2)$ chiral symmetry of continuum QCD (if two quarks are supposed to become massless), but only a $U(1)$ subgroup. The lack of chiral symmetry is apparent in the pion spectrum for staggered quarks, where there exists only a single pseudo-Goldstone pion, while the other pions acquire an additional mass from $\mathcal{O}(a^2)$ flavor mixing terms in the action. Therefore it is necessary to study QCD thermodynamics with a fermion discretization that does not suffer from lattice pathologies. In Sect. 2.4.3 we have encountered domain-wall fermions that even preserve chiral symmetry. We now show briefly recent domain-wall fermion calculations at finite temperature.

The domain-wall formalism is a variant of Wilson fermions (cf. Sect. 2.4.3) to the extent of a fifth dimension (the s direction). The left- and right-handed chiral states are bound to the four-dimensional boundaries of the five-dimensional volume. The finite separation, L_s, between the left- and right-hand boundaries allows then for some mixing between the left- and right-handed modes still giving rise to a residual chiral symmetry breaking. In contrast to Wilson fermions, this chiral symmetry breaking can be suppressed by taking L_s to be sufficiently large. Here we consider Ref. [76] where the region of the QCD phase transition is studied by using $2 + 1$ flavors of domain-wall fermions and a $16^3 \times 8$ lattice volume with a fifth dimension $L_s = 32$. The parameters in this calculation were chosen such that the pion mass is $m_\pi \approx 308\,\mathrm{MeV}$, more than twice the physical mass, and the kaon mass is $m_K \approx 496\,\mathrm{MeV}$, close to the physical one. The chiral order parameter is again given by the renormalized quantity $\Delta_{l,s}$ in Eq. (4.3.6), where the masses m_u, m_s are the bare masses and *not*, as one could expect, the "physical masses" $m_{\mathrm{res}} + m_u, m_{\mathrm{res}} + m_s$, cf. Sect. 2.4.3. The resulting chiral susceptibility (compare Eq. (4.3.7)) shown in Fig. 4.6 displays a clear peak around the lattice coupling strength $\beta_c = 2.03(1)$ and suggests a critical T-region between $155\,\mathrm{MeV}$ and $185\,\mathrm{MeV}$. From the peak position, a pseudo-critical temperature $T_c = 171(10)(17)\,\mathrm{MeV}$ is estimated in Ref. [76]. The first error represents the statistical and systematic errors when determining β_c and the corresponding physical scale at

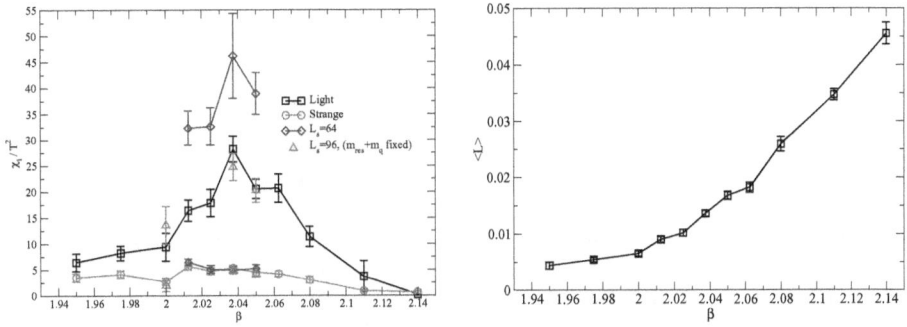

FIGURE 4.6: Left: β dependence of the chiral susceptibility χ_l/T^2 for light (black) and the strange (red) quark at $L_s = 32$; the result for the light-quark susceptibility at $L_s = 64$ is given by the blue diamonds. Both calculations were performed for *fixed* m_l and m_s (i.e., the physical mass $m_{res} + m_q$ is smaller for $L_s = 64$ because of a smaller residual mass m_{res}). The $L_s = 96$ calculation (pink triangles) was performed such as to reproduce the $L_s = 32$ value of the physical mass $m_{res} + m_q$. Right: β dependence of the Polyakov loop. (Plots from Ref. [76].)

larger than physical quark masses and nonzero lattice spacing. The second error is an estimate for the shift in T_c when lower quark masses are used and a continuum extrapolation is performed.

When comparing the chiral transition region to the rise of the Polyakov loop in Fig. 4.6 the authors of Ref. [76] draw the conclusion that there is evidence for a single crossover transition for both the chiral and the confinement-deconfinement transition. However, the Polyakov-loop susceptibility does not show any well-resolved peak in the data, therefore the authors of Ref. [76] were unable to locate the confinement-deconfinement crossover region using this variable. Finally, we have to comment on the impact of the residual quark mass m_{res} or, equivalently, on the size of L_s. The physically relevant quantity is the mass term $m_{res} + m_q$, where m_q is the bare quark mass. Hence, in particular the light quark mass is influenced by a variation of L_s. In order to investigate this impact, the simulations in Ref. [76] were performed at different values of L_s, on one hand, and by fixing $m_{res} + m_q$, on the other. As a conclusion of these calculations subject to the before-mentioned boundary conditions one extracts the critical value $\beta_c = 2.03(1)$. These simulations give first promising results for lattice calculations under "physical conditions". An improvement in computer power is, however, necessary in order to perform simulations using domain-wall fermions at physical pion masses.

We close this overview section by summarizing all results found and described here in Table 4.1.

4.4 Model Predictions

In the last two sections we discussed that a complete exploration of the QCD diagram is so far neither possible experimentally nor by lattice calculations. In a situation like this model calculations are certainly useful for further guidance. One of the most successful models used in the last decades, the Nambu–Jona-Lasinio model, has been presented in the first part of this

	$\chi_{m,l}/T^4$	$\chi_{m,l}/T^2$	$\chi_{m,l}$	$\Delta_{l,s}$	$L_{\rm ren}$
Ref. [95]	151(3)(3)	—	—	—	176(3)(4)
Ref. [77]	146(2)(3)	152(3)(3)	157(3)(3)	155(2)(3)	170(4)(3)
Ref. [99]	147(2)(3)	—	—	157(3)(3)	– – –
Ref. [75]	—	—	—	196(3)	196(3)
Ref. [96]	—	—	—	185–195	185–195
Ref. [76]	—	171(10)(17)	—	—	—

TABLE 4.1: Continuum extrapolated transition temperatures (in MeV) at the physical point for different observables and in different works. The uncertainty in the first parenthesis refers to $T > 0$, the second one to $T = 0$ statistical plus systematic errors. See the text for explanations.

work and will now be extended to the Polyakov–Nambu–Jona-Lasinio (PNJL) model suitable for the description of QCD thermodynamics. Many more models have been applied for this purpose. Some of the results for the location of the critical point are collected in Fig. 4.7, compiled by M. Stephanov in Ref. [103]. We can already state at this point that model calculations are very useful in order to get *qualitative* insights of the behavior of strongly interacting matter at high temperatures and densities. The *quantitative* results, however, depend strongly on the model and the parameters used. We will see this also in our work.

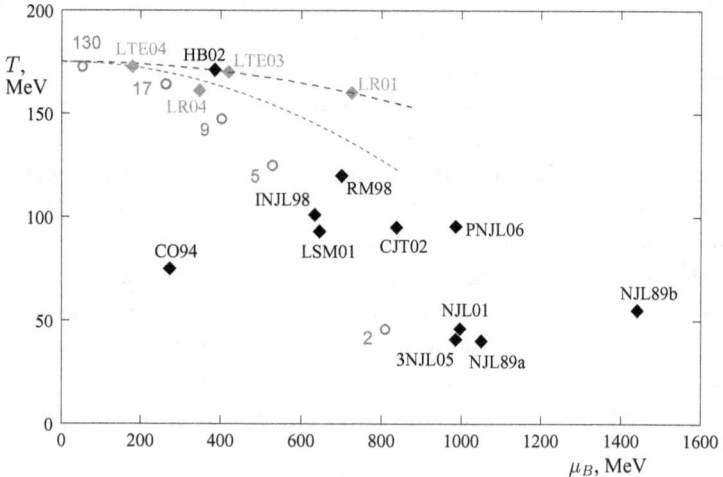

FIGURE 4.7: Comparison of predictions for the location of the QCD critical point from Ref. [103]. Black points are model predictions: NJLa89, NJLb89 [104] – CO94 [105–107] – INJL98 [108] – RM98 [109] – LSM01, NJL01 [110] – HB02 [111] – 3NJL05 [112] – PNJL06 [37]. Green points are lattice predictions: LR01, LR04 [89, 90] – LTE03 [113] – LTE04 [114]. The two dashed lines are parabolas with slopes corresponding to lattice predictions of the slope $dT/d\mu_B^2$ of the transition line at $\mu_B = 0$ [91, 92, 113]. The red circles are locations of the freeze-out points for heavy-ion collisions at corresponding center of mass energies per nucleon (indicated by labels in GeV).

5 Polyakov-Loop-Extended Nonlocal NJL Models

Here we focus on the finite-temperature and -density description of the two- and three-flavor nonlocal NJL models. Basic tools of QCD thermodynamics will first be summarized. We introduce the Polyakov loop as an order parameter for the confinement-deconfinement phase transition in pure gauge QCD without fermions. Of primary interest is the interrelation between the chiral and confinement transition. NJL models, in general, and our nonlocal framework, in particular, are able to describe spontaneous chiral symmetry breaking. They do not, however, incorporate confinement, because gluons are eliminated as dynamical degrees of freedom[1]. We explain the coupling of the Polyakov loop to the fermions and calculate the partition function describing the system. After the calculation in mean-field approximation at finite temperatures and densities we include second-order corrections to our finite-temperature calculations. Finally we present the phase diagram of two- and three-flavor QCD as a result of the nonlocal NJL calculations.

5.1 Thermodynamics of QCD

In thermal field theory (see, e. g., [8, 115, 116]) it is convenient to start from the path-integral formulation of quantum field theory. Consider the *partition function* of the grand-canonical ensemble, $\mathcal{Z}(\beta, \mu)$, defined through

$$\mathcal{Z}(\beta, \mu) = \mathrm{Tr}\left(\mathrm{e}^{-\beta(\hat{H} - \mu\hat{N})}\right),\tag{5.1.1}$$

where $\beta := 1/T$ is the inverse temperature, $\hat{H}$ the Hamiltonian describing the system, μ is the chemical potential and $\hat{N}$ the particle-number operator.[2] Once the partition function $\mathcal{Z}$ is

---

[1]This is obvious for the local NJL models. In the nonlocal approach, the situation is more intricate: it has often been argued (for a detailed exposure of the argument see Ref. [27]), that by an appropriate choice of the momentum distribution function $\mathcal{C}$, the fermion propagator (2.5.9) does not have purely imaginary poles in Euclidean space (corresponding to real poles in Minkowski space). From this, it is often argued that a decay into a pair of two constituent quarks is suppressed, which is considered a "form of" confinement.

[2]Following the statistical treatment of thermodynamics, the β and μ have to be considered initially as Lagrange multipliers that make sure that the averaged energy and particle number of the system remain constant. The relation to temperature and density is only given in thermodynamics by comparison with the laws of thermodynamics.

determined, one can derive easily the grand-canonical potential (density)

$$\Omega = -\frac{T}{V} \ln \mathcal{Z}(\beta, \mu), \tag{5.1.2}$$

where V is the volume of the system. From the maximum principle of the entropy it follows that the system always tends to minimize the thermodynamical potential Ω.

The partition function (5.1.1) can be expressed (cf. [115, 116]) as a functional integral in the same manner as the generating functional (2.3.3),

$$\mathcal{Z}(\beta, \mu) = \oint' \mathcal{D}\phi \exp\left(-\int_0^\beta d\tau \int d^3x \left(\mathcal{L}_E - \mu\hat{N}\right)\right), \tag{5.1.3}$$

where $\mathcal{L}_E$ denotes the Lagrangian expressed in Euclidean coordinates (see Eq. (2.5.6)), ϕ stands for an arbitrary field variable and $\oint'$ indicates that the fields ϕ are subject to periodic or antiperiodic boundary conditions for bosonic or fermionic fields, respectively. In Euclidean space-time we have:

$$\begin{aligned} \phi(\vec{x}, 0) &= \phi(\vec{x}, \beta), & \text{for } \textit{bosonic} \text{ fields } \phi \\ \psi(\vec{x}, 0) &= -\psi(\vec{x}, \beta), & \text{for } \textit{fermionic} \text{ field } \psi. \end{aligned} \tag{5.1.4}$$

These boundary conditions follow from the cyclicity of the trace as the so-called Kubo-Martin-Schwinger relations, which are a direct consequence of the canonical commutation and anticommutation relations for bosons and fermions. Bosonic fields have c-number eigenvalues, while fermionic fields have Grassmann-number eigenvalues. This implies that the expressions derived at zero temperature (Chapt. 3) can be used modifying the four-component p_4 of the momentum such that it meets the correct boundary condition. These replacement rules are given by the so-called *Matsubara formalism*.

Matsubara Formalism

In order to sketch the Matsubara formalism, consider the free bosonic and fermionic propagators[3] that fulfill periodic and antiperiodic boundary conditions, respectively, in Euclidean, i. e., imaginary time. Setting $\mu = 0$ and denoting the propagators by $D_\beta(\tau, \vec{x})$ and $S_F^\beta(\tau, \vec{x})$, respectively, it is always possible to write them as a Fourier series in imaginary-time space according to

$$D_\beta(\tau, \vec{x}) = \frac{1}{\beta} \sum_{n \in \mathbb{Z}} e^{-i\omega_n \tau} \hat{D}_\beta(i\omega_n, \vec{x}) \qquad \text{with} \quad \omega_n = \frac{2\pi n}{\beta} \tag{5.1.5}$$

$$S_F^\beta(\tau, \vec{x}) = \frac{1}{\beta} \sum_{n \in \mathbb{Z}} e^{-i\omega_n \tau} \hat{S}_F^\beta(i\omega_n, \vec{x}) \qquad \text{with} \quad \omega_n = \frac{(2n+1)\pi}{\beta}. \tag{5.1.6}$$

The frequencies ω_n appearing in both expressions are the so-called *Matsubara frequencies*. In order to establish the rules for using the Matsubara formalism at finite temperatures, we calculate

---

[3]The same results hold true for the general case, and we will use them in a more general framework. But the general derivation would require a discussion of spectral functions, which is beyond the scope of this work.

$\hat{D}_\beta$ and $\hat{S}_F^\beta$ in momentum space, leading to

$$\hat{D}_\beta(i\omega_n, \vec{k}) = \frac{1}{\omega_n^2 + \vec{k}^2 + m^2} \tag{5.1.7}$$

$$\hat{S}_F^\beta(i\omega_n, \vec{p}) = \frac{1}{\gamma_4 \omega_n - \vec{\gamma} \cdot \vec{p} + m}, \tag{5.1.8}$$

where we have used the standard *convention* for the Matsubara frequencies, such that $p_4 \to -\omega_n$ (cf. Eq. (2.5.9)). Before summarizing the relevant results, we have to come back shortly to the question of a finite chemical potential. Although true in general, we derive the replacement rule here for the fermionic case which is of relevance for the present work. For fermions one has the number operator $\hat{N} = \int d^4x\, \bar{\psi}(x)\gamma^0\psi(x) = \int d^4x\, \bar{\psi}(x)(-i\gamma_4)\psi(x)$. Inserting this explicitly in Eq. (5.1.3), one sees that the Euclidean time derivative, $\partial_4 = \partial_\tau$, has to be replaced according to $\partial_4 \to \partial_4 - \mu$. With $p_\mu = i\partial_\mu$ and $p_4 \to -\omega_n$ we obtain at finite chemical potential:

$$-p_4 \to \omega_n - i\mu. \tag{5.1.9}$$

This is true for both fermionic and bosonic *particles*. When dealing with *anti*particles, one has to replace $-i\mu \to +i\mu$.

Let us now summarize, how the zero-temperature description of Chapt. 3 can be extended to finite temperatures and densities: First, replace the four-component p_4 by the Matsubara frequencies ω_n, according to Eq. (5.1.9), and use the symmetric or antisymmetric expressions for ω_n in Eq. (5.1.5) depending on the treatment of bosons or fermions, respectively. Finally, replace the integral over the Euclidean space-time by

$$\text{(zero temperature)} \qquad \int \frac{d^4p}{(2\pi)^4} \quad \longrightarrow \quad \frac{1}{\beta} \sum_{\omega_n} \int \frac{d^3\vec{p}}{(2\pi)^3} \qquad \text{(finite temperature)}. \tag{5.1.10}$$

These prescriptions will be extensively applied in the following sections.

5.2 Polyakov Loop and Confinement-Deconfinement Transition

After the general introduction to thermodynamics of strongly interacting matter, we now address the question of how gluonic degrees of freedom can be introduced in the model in order to deal with the confinement-deconfinement transition. As already mentioned previously neither the local nor the nonlocal NJL models can account for this transition because of the absence of gluons as dynamical degrees of freedom. First, we consider the heavy-quark limit $m_q \to \infty$, i.e., the pure gauge case. In this limit the *Polyakov loop* serves as an exact order parameter for the confinement-deconfinement transition.

5.2.1 Polyakov Loop

Let us now introduce the *Wilson line* $\mathcal{W}_A$ encountered already in Sects. 2.4.3 and 3.2.2, cf. Eq. (2.4.15). It is a transformation operator connecting the configuration on the gauge group manifold of one point x in space-time to the configuration of another point y:[4]

$$\mathcal{W}_A(x,y) = \mathcal{P} \exp\left\{ i \int_x^y dx^\mu A_\mu(x) \right\} , \tag{5.2.1}$$

where $\mathcal{P}$ denotes the path-ordering operator. After an analytical continuation to the complex time plane, the Wilson line can be used to connect two points $\tau = 0$ and $\tau = \beta$ in imaginary space-time,

$$L = \mathcal{P} \exp\left\{ i \int_0^\beta d\tau\, A_4(\tau, \vec{x}) \right\} . \tag{5.2.2}$$

This is the so-called *Polyakov loop* [117]; in what follows, we are interested in the renormalized Polyakov loop Φ and its expectation value $\langle\Phi\rangle$ defined as[5]

$$\Phi(\vec{x}) = \frac{1}{N_c} \mathrm{Tr}_c \left[L(\vec{x}) \right] , \qquad \langle\Phi(\vec{x})\rangle = \frac{1}{N_c} \langle \mathrm{Tr}_c \left[L(\vec{x}) \right] \rangle . \tag{5.2.3}$$

We are now going to show (following Res. [97, 118]), that the Polyakov loop $\langle\Phi(\vec{x})\rangle$ can be related to the free energy $\mathscr{F}_{\bar{q}q}(\vec{x} - \vec{y}, T)$ at temperature T of two *static* color sources $\bar{q}$ and q with spatial separation $\vec{x} - \vec{y}$ according to

$$e^{-\beta \mathscr{F}_{\bar{q}q}(\vec{x}-\vec{y},T)} = \langle \Phi(\vec{x})\Phi^\dagger(\vec{y}) \rangle_\beta , \tag{5.2.4}$$

where $\langle\ \rangle_\beta$ denotes the thermal expectation value.

In order to prove Eq. (5.2.4) we start with the statistical definition of the free energy of a quark-antiquark pair, $\mathscr{F}_{q\bar{q}} = -\frac{1}{\beta} \ln \mathcal{Z}(\beta)$ with the *canonical* partition function $\mathcal{Z}(\beta) = \mathrm{Tr}\left(e^{-\beta \hat{H}} \right)$. Exponentiating this definition leads to

$$e^{-\beta \mathscr{F}_{q\bar{q}}(\vec{x},\vec{y})} = \frac{1}{N_c^2} \sum_{|s\rangle} \langle s | e^{-\beta \hat{H}} | s \rangle ,$$

with the summation indicated over all states $|s\rangle$ with a *heavy quark* at $\vec{x}$ and a *heavy antiquark* at $\vec{y}$. We are considering the heavy-quark limit, i. e., $m_q \to \infty$, because we are finally interested in the pure gauge case. The factor $1/N_c^2$ has been introduced for later convenience; since this is a constant added to the free energy, physics are not influenced by this. Next, we express the trace in the above formula in terms of states $|s'\rangle$ with *no* heavy quarks:

$$e^{-\beta \mathscr{F}_{q\bar{q}}(\vec{x},\vec{y})} = \frac{1}{N_c^2} \sum_{|s'\rangle} \left\langle s' \left| \sum_{a,b} \psi_b(\vec{y},0)^c \psi_a(\vec{x},0)\, e^{-\beta \hat{H}}\, \psi_a^\dagger(\vec{x},0) \psi_b^{\dagger c}(\vec{y},0) \right| s' \right\rangle .$$

The quark and charge-conjugate quark field operators ψ_a, ψ_a^c, respectively, (with color index a)

---

[4]From now on we absorb the coupling strength g in the gauge field, i. e., henceforth we write $A_\mu \to g A_\mu$.

[5]Note, that Eqs. (5.2.2) and (5.2.3) are the continuum expressions of Eq. (4.3.8) which defines L and $\langle\Phi\rangle$ on the lattice.

have been introduced here. The two given formulas for the free energy are, indeed, equivalent because of the equal-time anticommutation relations

$$\left\{\psi_a(\vec{x},t), \psi_b^\dagger(\vec{x}',t)\right\} = \delta^{(3)}(\vec{x}-\vec{x}')\delta_{ab}\,,$$

and similarly for ψ^c, with all other equal-time anticommutators vanishing. Therefore, the expectation value over s' gives only a contribution, if $\psi^\dagger, \psi^{\dagger c}$ are creation operators of *heavy* quarks.

Since $e^{-\beta\hat{H}}$ generates Euclidean time translations, i.e., $e^{\beta\hat{H}}\mathcal{O}(t)e^{-\beta\hat{H}} = \mathcal{O}(t+i\beta)$ for any operator $\mathcal{O}(t)$, we can now write (in Euclidean space-time):

$$e^{-\beta\mathscr{F}_{q\bar{q}}(\vec{x},\vec{y})} = \frac{1}{N_c^2}\sum_{|s'\rangle}\left\langle s'\left|\sum_{a,b}e^{-\beta\hat{H}}\,\psi_a(\vec{x},\beta)\psi_a^\dagger(\vec{x},0)\psi_b^c(\vec{y},\beta)\psi_b^{\dagger c}(\vec{y},0)\right|s'\right\rangle. \qquad (5.2.5)$$

Here we have used the equal-time anticommutation relation for the fermionic fields in combination with the antiperiodic boundary conditions (5.1.4) for the fields, leading to

$$\left\{\psi_a(\vec{x},\beta), \psi_b^\dagger(\vec{x}',0)\right\} = -\delta^{(3)}(\vec{x}-\vec{x}')\delta_{ab}\,.$$

Since we consider *static* quarks and antiquarks, they obey the static time-evolution equation (in Coulomb gauge)

$$\left(\frac{1}{i}\frac{\partial}{\partial t} - A_0(\vec{x},t)\right)\psi(\vec{x},t) = 0\,,$$

with $A_0 = t_a A_0^a$ and t_a being the generators of the $SU(N_c)$ Lie algebra. This equation can be integrated yielding

$$\psi(\vec{x},t) = \mathcal{T}\exp\left\{i\int_0^t dt'\, A_0(\vec{x},t')\right\}\psi(\vec{x},0)\,,$$

where $\mathcal{T}$ denotes the time-ordering operator. Inserting this solution, Eq. (5.2.5) can be expressed in terms of the Polyakov loop Φ, Eqs. (5.2.2) and (5.2.3),

$$e^{-\beta\mathscr{F}_{q\bar{q}}(\vec{x},\vec{y})} = \mathrm{Tr}\left[e^{-\beta\hat{H}}\Phi(\vec{x})\Phi^\dagger(\vec{y})\right]\,.$$

The trace is over states of pure gluon theory (i.e., for the heavy-quark limit) only. This is because the arising expectation values, e.g.,

$$\langle s'(\vec{x})|\psi(\vec{x},0)\psi^\dagger(\vec{x},0)|s'(\vec{x})\rangle = \langle s'(\vec{x})|\left(1 - \psi^\dagger(\vec{x},0)\psi(\vec{x},0)\right)|s'(\vec{x})\rangle$$

are zero, if $s'(\vec{x})$ is *not* a heavy-quark state, and 1, if $s'(\vec{x})$ *is* a heavy-quark state. Hence Eq. (5.2.4) follows.

Furthermore, owing to the cluster property of statistical mechanics, the correlations between $\Phi(\vec{x})$ and $\Phi^\dagger(\vec{y})$ vanish for $|\vec{x}-\vec{y}| \to \infty$, leading to

$$\mathscr{F}_\infty := \lim_{|\vec{x}-\vec{y}|\to\infty}\mathscr{F}_{\bar{q}q}(|\vec{x}-\vec{y}|,T) = -T\ln|\langle\Phi\rangle|^2\,. \qquad (5.2.6)$$

This expression can be related to the free energy of a single quark[6], $\mathscr{F}_\infty^q$, and antiquark, $\mathscr{F}_\infty^{\bar{q}}$, via

$$\langle \Phi \rangle = \mathrm{e}^{-\frac{1}{2}\beta\mathscr{F}_\infty^q(T)}, \qquad\qquad \langle \Phi^\dagger \rangle = \mathrm{e}^{-\frac{1}{2}\beta\mathscr{F}_\infty^{\bar{q}}(T)}. \qquad (5.2.7)$$

Consequently, $\langle \Phi \rangle$ is a measure for the confinement of the quarks while $\langle \Phi^\dagger \rangle$ indicates the strength of the confinement of the antiquarks. In particular, when considering the pure gauge case without quarks, the confined phase is characterized by an infinite free energy $\mathscr{F}$ which corresponds to a vanishing expectation value for the Polyakov loop, i.e., $\langle \Phi \rangle = 0$. On the other hand, if the system is deconfined, $\langle \Phi \rangle$ assumes a nonzero value.

We will now comment on the symmetry that underlies the confinement-deconfinement transition. In a purely gluonic system the confinement-deconfinement transition is connected to the spontaneous breaking of the center symmetry of SU(3). The center Z of a group includes all the elements of the group that commute with *all* group elements. Consequently, $Z(3) := Z(\mathrm{SU}(3)) = \{1, \mathrm{e}^{\frac{2}{3}\pi\mathrm{i}}, \mathrm{e}^{\frac{4}{3}\pi\mathrm{i}}\}$, which are the third roots of unity. Now let us consider the QCD Lagrangian (2.2.1) in the absence of quarks (or with static quarks):

$$\mathscr{L}_{\mathrm{gauge}} = -\frac{1}{2g^2}\,\mathrm{Tr}\left(G_{\mu\nu}G^{\mu\nu}\right). \qquad (5.2.8)$$

Next, calculate the partition function (cf. Eq. (5.1.3))

$$\mathcal{Z}_{\mathrm{gauge}} = \int \mathscr{D}A \,\exp\left(-\int_0^\beta \mathrm{d}\tau \int \mathrm{d}^3x\, \mathscr{L}_{\mathrm{gauge}}\right), \qquad (5.2.9)$$

with the fields A_μ subject to periodic boundary conditions (because of their bosonic character), i.e.,

$$A_\mu(\tau+\beta, \vec{x}) = A_\mu(\tau, \vec{x}). \qquad (5.2.10)$$

It turns out, that $\mathcal{Z}_{\mathrm{gauge}}$, subject to the previously mentioned boundary conditions, is invariant under local gauge transformations $g(x) = \mathrm{e}^{\mathrm{i}\,\xi_a(x)t_a}$ (with arbitrary real functions $\xi_a(x)$ and $gg^\dagger = 1$), if and only if the following condition is fulfilled:

$$g(\tau+\beta, \vec{x}) = zg(\tau, \vec{x}) \qquad \text{with } z = \mathrm{e}^{\frac{2\pi\mathrm{i}}{3}n}, \quad n \in \{1, 2, 3\}. \qquad (5.2.11)$$

It is clear that z is an element of $Z(3)$. Moreover, from this transformation behavior of the gauge fields one obtains the following transformation law for the Polyakov loop

$$\langle \Phi(\vec{x}) \rangle \xrightarrow{\mathrm{SU(3)_{local}}} \mathrm{e}^{\frac{2\pi\mathrm{i}}{3}n}\langle \Phi(\vec{x}) \rangle. \qquad (5.2.12)$$

We conclude that the Polyakov loop $\langle \Phi \rangle$ is invariant under $Z(3)$ transformations if it vanishes, which is, according to Eq. (5.2.7) equivalent to a confined system. On the other hand, for a deconfined system, characterized by $\langle \Phi \rangle \neq 0$, the Polyakov-loop variable is not invariant under $Z(3)$ transformations.

The Polyakov-loop variable $\langle \Phi \rangle$ is thus the order parameter for the confinement-deconfinement

---

[6]Note, though, that these quantities are physically meaningless.

phase transition (in the pure gluonic case): the confined phase is characterized by a restored center symmetry of the SU(3) Lie group, described by $\langle \Phi \rangle = 0$, and in the deconfined phase the $Z(3)$ symmetry is spontaneously broken, entailing $\langle \Phi \rangle \neq 0$ [119, 120] (see also Fig. 5.2).

The Polyakov loop $\langle \Phi \rangle$ serves as an exact order parameter only in the heavy-quark limit, $m_q \rightarrow \infty$. Consider the antiperiodic boundary conditions for fermionic fields, $\psi(\tau + \beta, \vec{x}) = -\psi(\tau, \vec{x})$. Under a local SU(3) gauge transformation g, these fields transform according to $\psi \rightarrow \psi' = g\psi$, compare Sect. 2.2. But this implies

$$\psi'(\tau + \beta, \vec{x}) = -z\psi'(\tau, \vec{x}) \,, \tag{5.2.13}$$

which is only possible for $z = 1$. Therefore, in the presence of quark fields, the center symmetry of the generating functional $\mathcal{Z}$ is explicitly broken. We will see in Sect. 5.3, that $\langle \Phi \rangle$ can nonetheless be used as an approximate order parameter for the confinement-deconfinement transition even if quarks are present, in the sense that $\langle \Phi \rangle$ changes rapidly at the transition.

5.2.2 The Effective Polyakov-Fukushima Potential

We show now how gluonic dynamics can be incorporated in NJL-type models using the Polyakov loop $\langle \Phi \rangle$. We follow Ref. [121] in choosing an appropriate ansatz for the effective potential that describes the gluonic dynamics. Consider the full QCD generating functional for the Euclidean action $\mathcal{S}_{\mathrm{E}}$ (cf. Eqs. (2.3.3') and (2.5.6)):

$$\int \mathscr{D}A \int \mathscr{D}\bar{\psi} \int \mathscr{D}\psi \, \mathrm{e}^{-\mathcal{S}_{\mathrm{E}}}$$
$$= \prod_{\omega, \vec{p}} \left\{ \int \cdots \int \left[\prod_{j \in \{1,2,4,5,6,7\}} \mathrm{d}A^j_{\omega, \vec{p}} \right] \int \mathrm{d}A^3_{\omega, \vec{p}} \int \mathrm{d}A^8_{\omega, \vec{p}} \int \mathrm{d}\bar{\psi}_{\omega, \vec{p}} \int \mathrm{d}\psi_{\omega, \vec{p}} \, \mathrm{e}^{-\mathcal{S}_{\mathrm{E}}} \right\} \,, \tag{5.2.14}$$

where the gluon fields[7] $A(t, \vec{x})$ are expressed in the basis of the Gell-Mann matrices λ_i,

$$A(t, \vec{x}) = \sum_{\omega, \vec{p}} \sum_{j=1}^{N_{\mathrm{c}}^2-1} A^j_{\omega, \vec{p}} \lambda_j \mathrm{e}^{\mathrm{i}(\omega t - \vec{p} \cdot \vec{x})} \,. \tag{5.2.15}$$

The N_{c}^2-1 gluon fields are separated into the set of nondiagonal matrices in the squared brackets and the diagonal *Cartan algebra* elements A^3 and A^8. We show in Appendix E (see Ref. [122]) that it is generally possible to parametrize the Lie group SU(N) such that its group volume measure element J (the *Haar measure*) depends exclusively on the diagonal elements of SU(N). Thus, the relevant physical content is already contained in the field components A^3 and A^8. This allows one, without loss of generality, to integrate out the nondiagonal matrices from the generating functional, leading to the integrated Haar-measure volume. For the case of SU(3) we obtain

$$J(\phi_3, \phi_8) = \frac{1}{V_{\mathrm{SU}(3)}} \int \prod_{j \in \{1,2,4,5,6,7\}} \mathrm{d}A^j = \frac{2}{3\pi^2} \left(\cos(\phi_3) - \cos(\sqrt{3}\phi_8) \right)^2 \sin^2(\phi_3) \,, \tag{5.2.16}$$

---

[7]Euclidean indices are suppressed for simplicity.

where we have set $\phi_{3,8} = \beta \frac{A_4^{3,8}}{2}$.

In the so-called Polyakov gauge, the Polyakov loop[8] Φ is parametrized only by the diagonal elements of the SU(3) Lie algebra,

$$\Phi = \frac{1}{N_c} \mathrm{tr}_c \left[\exp\left(\mathrm{i}(\phi_3 \lambda_3 + \phi_8 \lambda_8)\right)\right]. \tag{5.2.17}$$

Neglecting additionally spatial fluctuations of the fields, i. e.,

$$A_\mu = \delta_{\mu 4}(A_4^3 t_3 + A_4^8 t_8), \tag{5.2.18}$$

allows one to write the Haar measure (5.2.16) in terms of the Polyakov loop Φ and its complex conjugate Φ^*,

$$J(\Phi, \Phi^*) = \frac{9}{8\pi^2} \left[1 - 6\Phi^*\Phi + 4\left(\Phi^{*3} + \Phi^3\right) - 3\left(\Phi^*\Phi\right)^2\right]. \tag{5.2.19}$$

If this expression is rewritten as a potential term to the action in the generating functional (5.2.14), one arrives at the Polyakov-Fukushima effective potential $\mathcal{U}$:

$$\frac{\mathcal{U}(\Phi, \Phi^*, T)}{T^4} = -\frac{1}{2}b_2(T)\Phi^*\Phi + b_4(T) \ln\left[1 - 6\Phi^*\Phi + 4\left(\Phi^{*3} + \Phi^3\right) - 3(\Phi^*\Phi)^2\right], \tag{5.2.20}$$

where the temperature-dependent coefficients are given in parametrized form:

$$b_2(T) = a_0 + a_1\left(\frac{T_0}{T}\right) + a_2\left(\frac{T_0}{T}\right)^2 + a_3\left(\frac{T_0}{T}\right)^3$$
$$b_4(T) = b_4\left(\frac{T_0}{T}\right)^3.$$

It implicitly includes the effects of the six nondiagonal gluon fields [121, 123, 124]. The first term on the right-hand side is reminiscent of a Ginzburg-Landau ansatz. It is evident from Eq. (5.2.20), that $\mathcal{U}$ is invariant under $Z(3)$ transformations, but the cubic expressions in the logarithm break the U(1) symmetry: the effective potential does not introduce a higher symmetry.

The effective potential $\mathcal{U}$ is plotted in Fig. 5.2 for temperatures below and above a critical temperature T_0, exhibiting the spontaneous breaking of the center symmetry. The values of the coefficients, listed in Table 5.1, have first been determined in Ref. [37], fitting both the lattice data for pressure, entropy density and energy density, and the behavior of the Polyakov loop on the lattice (see Fig. 5.1). The temperature scale T_0 is identified with the critical temperature for the first-order (pure gauge) confinement-deconfinement phase transition. We choose in this work $T_0 = 270\,\mathrm{MeV}$ from Ref. [125], although it is possible, that the transition temperature may vary according to the number of active quark flavors, see Ref. [126].

We close this section by mentioning that in the original work by K. Fukushima, Ref. [123] and Refs. [121, 127], a two-parameter ansatz based on the strong-coupling limit has been used. This version of $\mathcal{U}$ differs from that used in the present work only at higher temperatures but both forms produce very similar pressure profiles at $T \lesssim 2T_c$, the temperature region of primary interest.

---

[8]From now on we omit the angled brackets in the definition of the (renormalized) Polyakov loop, i. e.,
$\langle \Phi \rangle \to \Phi$.

a_0	a_1	a_2	a_3	b_4
3.51	-2.56	15.2	-0.62	-1.68

TABLE 5.1: Parameters of the Polyakov potential $\mathcal{U}$ (from Ref. [37]).

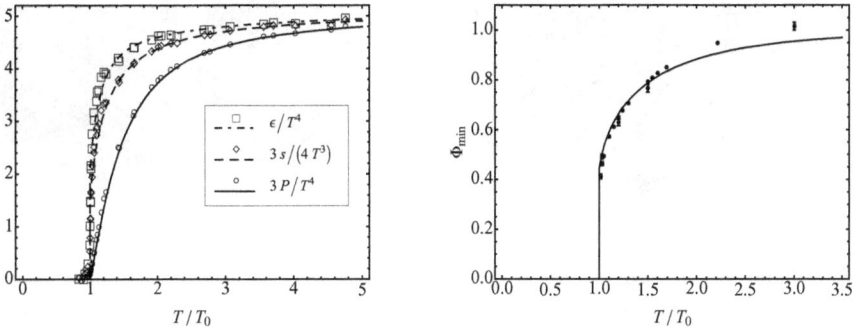

FIGURE 5.1: Simultaneous fit [37] of the pressure P, entropy density s and energy density ϵ (left), and of the Polyakov loop $\Phi_{\min}$ (right) calculated from the Polyakov potential $\mathcal{U}$. Lattice data from Refs. [128] and [129].

At higher temperatures, transverse gluon degrees of freedom—not covered by the Polyakov loop directly—begin to be important. For a discussion of transverse gluons in the nonlocal PNJL model see also Sect. 6.2.1.

5.3 Nonlocal Polyakov-Loop-Extended NJL Models

In the previous two sections we have introduced the necessary ingredients for a thermodynamic description of the NJL models. In the following, we apply the Matsubara formalism to the two- and three-flavor nonlocal NJL models and introduce the effective Polyakov-Fukushima potential (5.2.20) together with the Polyakov loop Φ as an (approximate) order parameter for the confinement-deconfinement transition.

5.3.1 Coupling of the Polyakov Loop to the Quarks

The coupling of the Polyakov loop Φ to the quark fields in the nonlocal NJL models is accomplished by the minimal gauge coupling procedure applied to nonlocal field theories. This implies, on one hand, a replacement of the partial derivative ∂_μ by a covariant derivative $D_\mu = \partial_\mu - iA_\mu$, or in momentum space $p_\mu \to p_\mu + A_\mu$.[9] Furthermore, one introduces the gluonic fields writing a Wilson line, Eq. (5.2.1), between the (nonlocal) fermionic bilinears, i.e., $\bar{\psi}(x)\psi(y) \to \bar{\psi}(x)\mathcal{W}_A(x,y)\psi(y)$. In the present case only the constant fields A_4^3 and A_4^8 enter, so the Wilson line simplifies to a phase factor $e^{i(x_4-y_4)(A_4^3 t_3 + A_4^8 t_8)}$. Thus, the minimal coupling pro-

---

[9]The coupling strength g is absorbed in the definition of the fields A_μ.

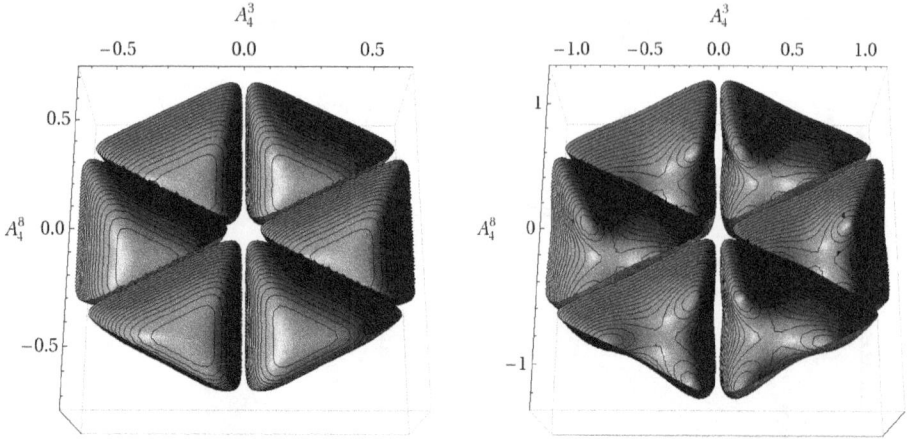

FIGURE 5.2: Effective Polyakov-Fukushima potential $\mathcal{U}$ shown as a function of A_4^3 and A_4^8. The left picture was generated for a temperature $T < T_0 = 270\,\text{MeV}$: in each of the triangular sections there is only one minimum that is invariant under $Z(3)$ transformations, hence describing the confined phase. In the right picture one has $T > T_0$ and in each compartment there are three degenerate minima, responsible for the spontaneous breakdown of the center symmetry and describing the deconfined phase. The overall hexagonal structure of both plots stems from the $Z(3)$ symmetry of the potential $\mathcal{U}$ itself.

cedure leads to a modification of the 4-component of the Euclidean momentum. In momentum space, the phase factor translates into delta functions, with the replacements

$$\vec{p} \to \vec{p}, \qquad p_4 \to p_4 + (A_4^3 t_3 + A_4^8 t_8)\,. \tag{5.3.1}$$

From Eq. (5.1.9) it follows, how to promote this to the respective finite-temperature formulations.

Mean-Field Approximation

When calculating the partition function $\mathcal{Z}(\beta, \mu)$, Eq. (5.1.3), in mean-field approximation (see also Sects. 3.2.2 and 3.3.3), the field variables ϕ are replaced by constants $\bar{\phi}$ that are determined by the principle of least action. In this case the partition function is given by an expression $\bar{\mathcal{Z}} = \mathrm{e}^{-\bar{\mathcal{S}}_\mathrm{E}}$, where the bars denote quantities evaluated at the mean-field configuration $\bar{\phi}$. From Eq. (5.1.2) the following relation results between the mean-field action $\bar{\mathcal{S}}_\mathrm{E}$ and the thermodynamic potential Ω:

$$\bar{\Omega} = \frac{T}{V}\bar{\mathcal{S}}_\mathrm{E}\,. \tag{5.3.2}$$

This relation is useful because it allows one to determine the thermodynamical potential in mean-field approximation by using the expressions for the Euclidean action derived at zero temperature and applying the Matsubara formalism. The mean fields, in turn, are determined by the stationary condition of the grand-canonical potential, i. e.,

$$\frac{\partial \Omega}{\partial \phi} = 0 \qquad \text{for } \phi = \bar{\phi}. \tag{5.3.3}$$

Here, ϕ stands for an arbitrary field, in our case either a boson field or the Polyakov-loop variables A_4^3, A_4^8.

5.3.2 The Two-Flavor Nonlocal PNJL Model

We now proceed to evaluate the partition function (5.3.2) in mean-field approximation, first for the two-flavor model [39]. Later in this section we will also include corrections to the mean-field results. The three-flavor case is treated in the next section.

We start from the action at zero temperature, Eq. (3.2.13), and use the replacement rules (5.3.1), (5.1.9) and (5.1.10) in order to obtain the following expression for the thermodynamical potential:[10]

$$\Omega = -\frac{T}{2} \sum_{n \in \mathbb{Z}} \int \frac{\mathrm{d}^3 p}{(2\pi)^3} \, \mathrm{Tr} \ln \left[\beta \tilde{S}^{-1}(\mathrm{i}\omega_n, \vec{p}) \right] + \frac{\bar{\sigma}^2}{2G} + \mathcal{U}(\Phi, \Phi^*, T) . \tag{5.3.4}$$

Here $\tilde{S}^{-1}$ is the inverse quark propagator expressed in Nambu-Gor'kov space,

$$\tilde{S}^{-1}(\mathrm{i}\omega_n, \vec{p}) = \begin{pmatrix} \mathrm{i}\omega_n \gamma_0 - \vec{\gamma} \cdot \vec{p} - \hat{M} - \mathrm{i}(A_4 + \mathrm{i}\mu)\gamma_0 & 0 \\ 0 & \mathrm{i}\omega_n \gamma_0 - \vec{\gamma} \cdot \vec{p} - \hat{M}^* + \mathrm{i}(A_4 + \mathrm{i}\mu)\gamma_0 \end{pmatrix}, \tag{5.3.5}$$

where the momentum-dependent mass matrix $\hat{M}$ is diagonal in color space,

$$\hat{M} = \mathrm{diag}_c(M(\omega_n^-, \vec{p}), M(\omega_n^+, \vec{p}), M(\omega_n^0, \vec{p})) \tag{5.3.6a}$$

with

$$\omega_n^{\pm} = \omega_n - \mathrm{i}\mu \pm A_4^3/2 - A_4^8/(2\sqrt{3}), \quad \omega_n^0 = \omega_n - \mathrm{i}\mu + A_4^8/\sqrt{3} \tag{5.3.6b}$$

and $\omega_n = (2n+1)\pi T, n \in \mathbb{Z}$ are the fermionic Matsubara frequencies. $M(p_4, \vec{p})$ is the dynamically generated mass, Eq. (3.2.15a), already encountered in the zero-temperature treatment.[11] The trace can be further simplified leading to

$$\Omega = -4T \sum_{i=0,\pm} \sum_{n \in \mathbb{Z}} \int \frac{\mathrm{d}^3 p}{(2\pi)^3} \ln \left[{\omega_n^i}^2 + \vec{p}^2 + M^2(\omega_n^i, \vec{p}) \right] + \frac{\bar{\sigma}^2}{2G} + \mathcal{U}(\Phi, \Phi^*, T) . \tag{5.3.4'}$$

This is the thermodynamic potential of the two-flavor nonlocal PNJL model in mean-field approximation.

---

[10]Note an extra factor $1/2$ because of the doubling of the degrees of freedom in Nambu-Gor'kov space. This notation is useful for a separate treatment of particle and antiparticles.

[11]From first principles, the matrix $\hat{M}$ contains different contributions from the up and the down quark. Since we consider here only the isospin limit, $\hat{M}$ is proportional to unity in (two-)flavor space. In the three-flavor case described below, we will have to abandon this assumption owing to the considerable mass difference between the up and the strange quark.

Gap Equations and Results

We restrict ourself first to the zero-density case, i. e., we set $\mu = 0$ unless otherwise stated. Finite density will be introduced in an extra section.

The gap equations (5.3.3) are derived taking variations with respect to the sigma field and the Polyakov-loop variables Φ and Φ^*, or A_4^3 and A_4^8, that mimic the thermodynamics of confinement and deconfinement. The necessary conditions are given by the requirement of a stationary potential:

$$\frac{\partial \Omega}{\partial \bar{\sigma}} = \frac{\partial \Omega}{\partial A_4^3} = \frac{\partial \Omega}{\partial A_4^8} = 0 \, . \tag{5.3.7}$$

According to Refs. [37, 38], in mean-field approximation one has $\Phi = \Phi^*$ and, consequently, $A_4^8 = 0$.

We do not write down the resulting gap equations explicitly, because they can only be solved numerically. A consistency check is, however, to compare the gap equation for the σ field to the corresponding equation (3.2.18) at zero temperature. Performing the substitutions dictated by the Matsubara formalism, we indeed obtain the finite-temperature result. In Fig. 5.3 we show the results for the temperature dependence of the chiral condensate and of the Polyakov loop using the parameters given in Tables 3.1, 5.1. This figure shows clearly the entanglement of chiral dynamics and Polyakov-loop degrees of freedom (solid lines). As indicated by the dashed curves, without a coupling between quark quasiparticles and Polyakov loop (and in the chiral limit), the second-order chiral phase transition and the first-order deconfinement transition (of pure gauge QCD) appear at very different critical temperatures ($T_{\text{chiral}} \approx 110\,\text{MeV}$ for the chiral phase transition and $T_0 \approx 270\,\text{MeV}$ for deconfinement). The quark coupling to the Polyakov loop moves the deconfinement transition to lower temperature. At the same time the chiral transition (with explicit symmetry breaking by nonzero quark mass) turns into a crossover at an upward-shifted temperature, just so that both transitions nearly coincide in a common temperature range around 200 MeV. Note that the Polyakov loop in the presence of quarks is not a strict order parameter anymore. Usually one uses the point of steepest ascent of Φ in order to determine the transition temperature for the confinement-deconfinement transition, and the point of the steepest descent for the chiral condensate.

There is no *a priori* reason that dictates the coincidence of the two transitions. It was shown by G. 't Hooft [130] by general symmetry arguments that confinement always implies chiral symmetry breaking at zero temperature and zero baryon density. A simple and transparent argument was given by A. Casher [131] which also holds at finite temperatures[12]. Casher's argument uses the fact that a confined quark-antiquark bound state, say, is formed by superposing paths in which the bound fermion has to reverse its direction of motion at some point. Since its helicity (chirality) is fixed, the change of direction of motion implies a spin flip $\Delta S = 1$. Since the angular momentum is conserved, and the orbital angular momentum does not change, such a spin flip is not allowed, assuming a spin-independent attraction. Therefore, the helicity changes at the turning point. Thus, confinement of quarks indeed requires dynamical breaking of chiral

---

[12]Note, that Casher's argument does not exclude, however, the existence of chirally symmetric hadrons at large density [132].

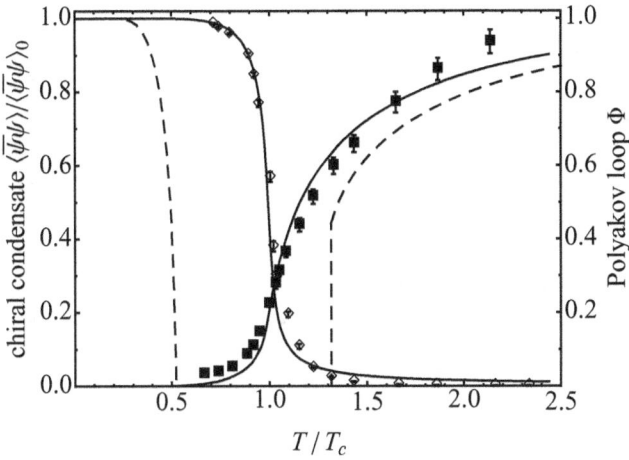

FIGURE 5.3: Solid curves: calculated temperature dependence of the chiral condensate $\langle\bar{\psi}\psi\rangle$ and of the Polyakov loop Φ normalized to the critical temperature $T_c = 207\,\mathrm{MeV}$ as obtained in the two-flavor nonlocal PNJL model. Also shown for orientation are lattice results from Ref. [133]. The dashed lines show the chiral condensate in the chiral limit (left for the pure fermionic case) and of the Polyakov loop for the pure gluonic case, respectively.

symmetry.

Comparing our two-flavor model results to lattice-QCD calculations [133] we observe that a similar symmetry breaking pattern is realized there, leading to a common chiral and deconfinement transition temperature $T_c = (196 \pm 3)\,\mathrm{MeV}$ (see, however, the discussion in Sect. 4.3.2). Fig. 5.3 includes these lattice data for orientation; a direct comparison with two-flavor nonlocal PNJL calculations is, however, not possible since the lattice computation has been performed with $N_\mathrm{f} = 2+1$ flavors and a pion mass $m_\pi \simeq 220\,\mathrm{MeV}$. A comparison to our three-flavor results presented in Sect. 5.3.3 will, hence, be more meaningful.

Pressure and Mesonic Corrections

From general thermodynamical considerations one obtains the pressure P from the grand-canonical potential Ω by[13]

$$P = -\Omega\,. \tag{5.3.8}$$

Once the pressure is calculated, all thermodynamically relevant quantities, such as the energy density, can be derived.

So far the calculations have been performed in the mean-field approximation in which the pressure P is determined by the quarks moving as quasiparticles in the background provided

---

[13]Note, that from Eq. (5.1.2) it follows that in our notation the grand-canonical potential has already been divided by the volume.

by the expectation values of the sigma field, $\bar{\sigma}$, and the Polyakov loop Φ. In order to get a realistic description of the hadronic phase (at temperatures $T \lesssim T_c$), it is important to include mesonic correlations. The hadronic phase is dominated by pions as Goldstone bosons and their interactions. The quark-antiquark continuum is suppressed by confinement. However, as pointed out in Ref. [134], mesons described within the standard (local) PNJL model can still undergo unphysical decays into the quark-antiquark continuum even below T_c. This is simply due to the fact that the fermion propagator (2.5.9) has purely imaginary poles in Euclidean space, or real poles in Minkowski space. In the nonlocal PNJL model, those unphysical decays do not necessarily appear by virtue of the momentum-dependent dynamical quark mass (see also the discussion in Sect. 2.5 and Ref. [27]). Indeed, we find no purely imaginary poles in the fermion propagator and hence mesons are stable well below T_c. This means that the pressure below T_c is basically generated by the pion pole with its almost temperature-independent position. Therefore, the calculated pressure below T_c corresponds to that of a boson (pion) gas with constant mass.

To include the mesonic contributions to the pressure in our nonlocal PNJL model we apply again the Matsubara formalism. The contribution of the pion and sigma to the pressure is given by their inverse propagators (3.2.17)[14] involving the quark loop contribution to the mesonic self-energies $\Pi_{\pi,\sigma}(\nu_m, \vec{p})$ (where $\nu_m = 2\pi m T, m \in \mathbb{Z}$ is the bosonic Matsubara frequency and $\vec{p}$ is the momentum of the incoming pion or sigma), depicted in Eq. (3.2.17), at finite temperature. One finds:

$$\Pi_{\pi,\sigma}(\nu_m, \vec{p}) = 8T \sum_{i=0,\pm} \sum_{n \in \mathbb{Z}} \int \frac{\mathrm{d}^3 k}{(2\pi)^3} \, \mathcal{C}(\omega_n^i, \vec{k} + \vec{p})^2 \times$$

$$\times \frac{\left(\omega_n^i + \frac{\nu_m}{2}\right)\left(\omega_n^i - \frac{\nu_m}{2}\right) + \left(\vec{k} + \frac{\vec{p}}{2}\right)\left(\vec{k} - \frac{\vec{p}}{2}\right) \pm M\left(\omega_n^i + \frac{\nu_m}{2}, \vec{k} + \frac{\vec{p}}{2}\right) M\left(\omega_n^i - \frac{\nu_m}{2}, \vec{k} - \frac{\vec{p}}{2}\right)}{\left[\left(\omega_n^i + \frac{\nu_m}{2}\right)^2 + \left(\vec{k} + \frac{\vec{p}}{2}\right)^2 + M^2\left(\omega_n^i + \frac{\nu_m}{2}, \vec{k} + \frac{\vec{p}}{2}\right)\right]\left[\left(\omega_n^i - \frac{\nu_m}{2}\right)^2 + \left(\vec{k} - \frac{\vec{p}}{2}\right)^2 + M^2\left(\omega_n^i - \frac{\nu_m}{2}, \vec{k} - \frac{\vec{p}}{2}\right)\right]},$$
$$(5.3.9)$$

with $\omega_n^\pm = \omega_n \pm A_4^3, \omega^0 = \omega_n$ and $M(\omega_n, \vec{p}) = m_q + \mathcal{C}(\omega_n, \vec{p}) \, \bar{\sigma}$. The additional contribution of mesonic quark-antiquark modes to the pressure calculated from the inverse meson propagators (evaluated at the mean-field values $\sigma = \bar{\sigma}$ and $\pi = \bar{\pi} = 0$) represents a ring sum of random phase approximation chains, investigated in Ref. [135], and leading to the expression

$$P_{\mathrm{meson}}(T) = -T \sum_{M=\pi,\sigma} \frac{d_M}{2} \sum_{m \in \mathbb{Z}} \int \frac{\mathrm{d}^3 p}{(2\pi)^3} \ln\left[1 - G\Pi_M(\nu_m, \vec{p})\right], \qquad (5.3.10)$$

where d_M is the mesonic degeneracy factor, i.e., $d_\pi = 3, d_\sigma = 1$. Due to the momentum dependence of the nonlocality distribution $\mathcal{C}(p)$ and the dynamical quark mass $M(p)$, integrations and summations in Eqs. (5.3.9) and (5.3.10) can only be carried out numerically.

Results for the pressure in the presence of pion and sigma mesonic modes are presented in Fig. 5.4. Apart from the full result (solid line) we additionally show the mean-field result (with the pressure determined by quark quasiparticles only) and the mean-field result plus pion con-

---

[14]The same formula as in Eq. (5.3.4) applies with the fermion propagator $\tilde{S}^{-1}$ replaced by the inverse pion and sigma propagators F^- and F^+ from Eq. (3.2.17), respectively.

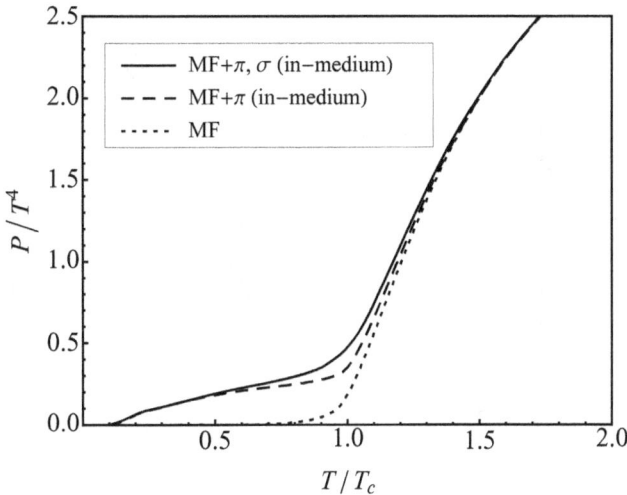

FIGURE 5.4: Pressure in units of T^4 calculated in the two-flavor nonlocal PNJL model as a function of temperature normalized to the critical temperature. Solid curve: full calculation (i. e., mean-field result plus mesonic corrections). Dotted curve: mean-field result (no mesonic corrections). Dashed curve: mean field plus pionic modes (no sigma).

tributions. The full pressure shows, indeed, the behavior discussed above: at low temperatures the mean-field contribution from quark quasiparticles is suppressed and the pressure can be described by a free pion gas. Near the critical temperature the sigma mesonic mode gives a small additional contribution. Above temperatures $T > 1.5\,T_c$ the mesonic contributions become negligible and the quark-gluon mean fields dominate the pressure.

Finally, in Fig. 5.5 we show a comparison of the pressure calculated with the physical pion mass, $m_\pi = 140\,\text{MeV}$, and with a "heavy" pion ($m_\pi = 500\,\text{MeV}$) corresponding to a quark mass of order $m_q \sim 100\,\text{MeV}$ that has frequently been used in earlier lattice-QCD computations. In this case the mesonic contributions to the pressure are evidently reduced. This explains the apparent agreement of lattice data with mean-field calculations [37, 67].

Related Thermodynamic Quantities

From the thermodynamical potential Ω further quantities of interest can be derived, such as the energy density ϵ, the trace anomaly $(\epsilon - 3P)/T^4$ and the sound velocity v_s. The trace anomaly is of particular interest as this is the quantity which is directly computed in lattice simulations[15] (Ref. [133]). A comparison to lattice data can serve only for orientation at this point. Since calculations in Ref. [133] were performed for $2+1$ flavors, it is more sensible to postpone a more profound comparison and discussion to Sect. 5.3.3, where the impact of the strange quark and

---

[15]Refs. [75, 96] induce the location of the transition region by studying the trace anomaly $(\epsilon - 3P)/T^4$ and find the same results as stated in Sect. 4.3.2 for the chiral order parameter $\Delta_{l,s}$ calculated in those references.

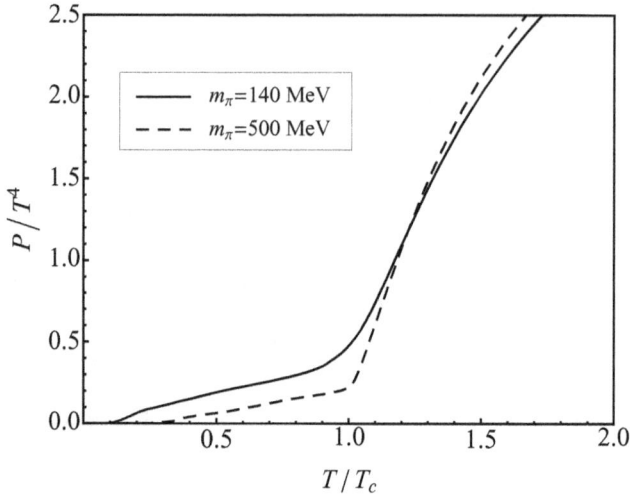

FIGURE 5.5: Comparison of the pressure for physical pion mass, $m_\pi = 140\,\text{MeV}$ (solid line), and heavy pion mass, $m_\pi = 500\,\text{MeV}$ (dashed line). Note that the crossing of the curves is due to the different critical temperatures to which they have been normalized.

the axial anomaly is additionally considered. For the time being, we limit ourself basically to the investigation of higher-order corrections to the mean-field results.

Consider first the trace anomaly related to the trace of the energy-momentum tensor. From the gauge action in $4 - 2\varepsilon$ dimensional Euclidean space-time,

$$S_{\text{E}}^{\text{gauge}} = \int \mathrm{d}\tau \int \mathrm{d}^{3-2\varepsilon}\vec{x} \left\{ \frac{1}{4g_0^2} G_{\mu\nu}^a G_{\mu\nu}^a \right\} ,$$

one calculates the Euclidean energy-momentum tensor

$$T_{\mu\nu} = \frac{1}{g_0^2} \left(\frac{1}{4}\delta_{\mu\nu} G_{\alpha\beta}^a G_{\alpha\beta}^a - G_{\alpha\mu}^a G_{\alpha\nu}^a \right) .$$

Here we have adopted the lattice convention of the action writing the (bare) coupling constant g_0 explicitly in front of the squared field strength tensor $G_{\mu\nu}^a$. The trace of the energy-momentum tensor in $\delta_{\mu\mu} = 4 - 2\varepsilon$ dimensions is then

$$T_{\mu\mu} = -\frac{2\varepsilon}{4}\frac{1}{g_0^2} G_{\alpha\beta}^a G_{\alpha\beta}^a .$$

Expressing the bare coupling strength g_0 in terms of the renormalized quantity g according to $g_0^2 = g^2 - \frac{11N_c}{3\varepsilon}\frac{g^4}{(4\pi)^2} + \mathcal{O}(g^6)$, we finally obtain

$$T_{\mu\mu} = -\frac{11N_c}{6}\frac{1}{(4\pi)^2} G_{\alpha\beta}^a G_{\alpha\beta}^a + \mathcal{O}(g^2) ,$$

which is the so-called trace anomaly.

Assuming a homogeneous and isotropic medium, the (Euclidean) energy-momentum tensor is given as $T_{\mu\nu} = \mathrm{diag}(\epsilon, -P, -P, -P)$ where ϵ is the energy density and P the pressure. The trace anomaly $(\epsilon - 3P)/T^4$ is evidently a direct measure of the deviation of the equation of state of QCD matter from that of a noninteracting, homogeneous and isotropic system. A major contribution to $\epsilon - 3P$ is expected to come from the gluon condensate. In the presence of quarks with nonvanishing masses, additional terms come from quark condensates. The impact of the trace anomaly, or: interaction measure, is expected to be maximal in the (phase) transition region. Finally, using standard thermodynamical relations, the trace anomaly is expressed in terms of a derivative of the pressure with respect to temperature according to

$$\frac{\epsilon - 3P}{T^4} = T\frac{\partial}{\partial T}\left(\frac{P}{T^4}\right). \tag{5.3.11}$$

The result is shown in Fig. 5.6. Once having determined the trace anomaly and the pressure, one deduces easily from Eq. (5.3.11) the energy density ϵ and the ratio of pressure and energy density, which enables us finally, to determine the square of the sound velocity (at constant entropy S):

$$v_s^2 = \left.\frac{\partial P}{\partial\epsilon}\right|_S = \frac{\left.\frac{\partial P}{\partial T}\right|_V}{T\left.\frac{\partial^2 P}{\partial T^2}\right|_V},$$

The following figures show the quantities just mentioned as they result in the mean-field case and with inclusion of mesonic corrections. Again, mesonic corrections are important only at temperatures below T_c (Figs. 5.6–5.9). Mesonic correlations do have a strong influence on the sound velocity below T_c (Fig. 5.9).

Finite Quark Chemical Potential and Phase Diagram

In this subsection we are going to extend the nonlocal PNJL model to finite quark chemical potential μ. We do this here with the aim of drawing a schematic phase diagram in the (T, μ) plane. For the sake of simplicity we restrict ourself to a scenario without diquarks, i.e., we do not describe a possible color-superconducting phase. We have set the stage for the finite-density description already at the beginning of this section. That means we calculate the thermodynamical potential Ω, Eq. (5.3.4), using the fermion propagator (5.3.5) where we now maintain a nonzero chemical potential. In the nonlocal PNJL model the four-component of the momentum appearing in the dynamical quark mass $M(p)$ experiences a modification due to both a nonzero μ and the Polyakov-loop variables A_4^3, A_4^8, according to Eqs. (5.1.9) and (5.3.1). In isospin symmetric matter, no distinction is made between an up- and down-quark chemical potential: we have $\mu_u = \mu_d = \mu$ that is related to the baryon chemical potential μ_B via $\mu_B = 3\mu$.

In the mean-field approximation we have $A_4^8 = 0$ [36] and determine the temperature and density dependence of the chiral order parameter, the σ field, by solving the gap equations (5.3.7) for the T- and μ-dependent thermodynamical potential Ω. The result is shown in Fig. 5.10. The profile of $\bar{\sigma}$ displays once again the chiral crossover transition at $\mu = 0$. It turns into a first-order phase transition at a critical point (here: $T_{\mathrm{CEP}} = 167\,\mathrm{MeV}$ and $\mu_{\mathrm{CEP}} = 175\,\mathrm{MeV}$). This

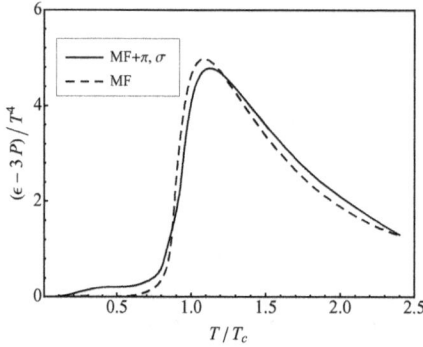

FIGURE 5.6: Trace anomaly $(\epsilon - 3P)/T^4$ shown as a function of temperature for the mean-field case (MF) and with mesonic correlations added.

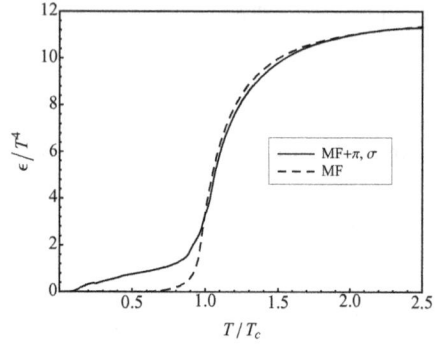

FIGURE 5.7: Comparison of the energy density ϵ as a function of temperature for the mean-field case and with mesonic correlations.

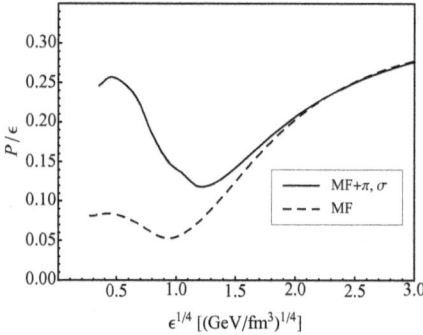

FIGURE 5.8: Fraction of pressure and energy density P/ϵ as a function of fourth root of the energy density $\epsilon^{1/4}$.

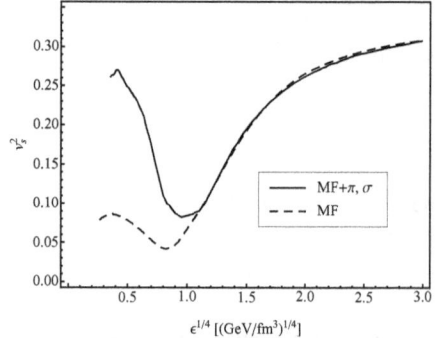

FIGURE 5.9: Squared sound velocity v_s^2 as a function of fourth root of the energy density $\epsilon^{1/4}$.

qualitative feature is generic for NJL- or PNJL-type models with or without explicit diquark degrees of freedom (see, e. g., Refs. [37, 67, 136]).

Of course, the two-flavor picture shown in Fig. 5.10 is quite schematic. A more "realistic" picture will be given in the next section 5.3.3, where strangeness is included. Even then, we do not get a well controlled location for the critical point. It turns out that the position of the critical point in the T-μ plane is sensitive to the model parameters. Furthermore, the almost constant behavior of $\bar{\sigma}(T = 0, \mu)$ with increasing chemical potential is unrealistic in the absence of explicit baryon (nucleon) degrees of freedom including their interactions. We will comment on these issues more profoundly after the treatment of the thermodynamic description of the three-flavor nonlocal model.

Irrespective of the simplifications made and the missing degrees of freedom at high densities, the nonlocal PNJL approach is useful and instructive in modeling the chiral and deconfinement

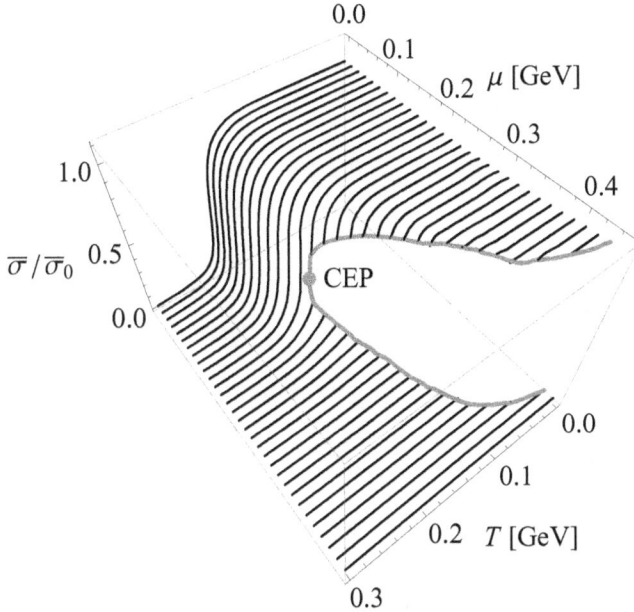

FIGURE 5.10: (Normalized) chiral order parameter $\bar{\sigma}/\bar{\sigma}_0$ shown in the (T, μ) plane. For small chemical potentials a crossover transition is manifest. For large chemical potentials a first-order phase transition is apparent which terminates in the critical point (CEP) at $(T_{CEP}, \mu_{CEP}) = (167\,\text{MeV}, 175\,\text{MeV})$.

thermodynamics at $\mu = 0$. The extension to the three-flavor nonlocal PNJL model leads to further interesting insights. We can already state the major improvement of the nonlocal approach compared to the local one: since no cutoff enters the calculations, the applicability of the nonlocal models is not restricted to a limited range of temperatures or densities. The question to be tackled is, however, which degrees of freedom become important when going to high chemical potentials.

5.3.3 Three-Flavor Nonlocal PNJL Model

The finite-temperature and -density formulation for three quark flavors is more intricate owing to the different up-/down- and strange-quark masses, on one hand, and the axial anomaly, on the other [40].

In mean-field and stationary phase approximation we can calculate the thermodynamical potential, like in the two-flavor case, from the action $\mathcal{S}_E^{MF}$, Eq. (3.3.29), taking into account the stationary phase approximation equations (3.3.31a). In the Matsubara formalism the thermodynamical (grand-canonical) potential reads[16]:

$$\Omega = -\frac{T}{2} \sum_{n \in \mathbb{Z}} \int \frac{\mathrm{d}^3 p}{(2\pi)^3} \, \mathrm{tr} \, \ln \left[\beta \tilde{S}^{-1}(\mathrm{i}\omega_n, \vec{p}) \right]$$
$$- \frac{1}{2} \left\{ \sum_{f \in \{u,d,s\}} \left(\bar{\sigma}_f \bar{S}_f + \frac{G}{2} \bar{S}_f \bar{S}_f \right) + \frac{H}{2} \bar{S}_u \bar{S}_d \bar{S}_s \right\} + \mathcal{U}(\Phi, \Phi^*, T), \tag{5.3.12}$$

with

$$\tilde{S}^{-1}(\mathrm{i}\omega_n, \vec{p}) = \begin{pmatrix} \mathrm{i}\omega_n \gamma_0 - \vec{\gamma} \cdot \vec{p} - \hat{M} - \mathrm{i}(A_4 + \mathrm{i}\hat{\mu})\gamma_0 & 0 \\ 0 & \mathrm{i}\omega_n \gamma_0 - \vec{\gamma} \cdot \vec{p} - \hat{M}^* + \mathrm{i}(A_4 + \mathrm{i}\hat{\mu})\gamma_0 \end{pmatrix} \tag{5.3.13}$$

where the momentum-dependent dynamical mass matrix $\hat{M}$ is diagonal in color and flavor space,

$$\hat{M} = \begin{pmatrix} \mathrm{diag}_c(M(\omega_{u,n}^-, \vec{p}), M(\omega_{u,n}^+, \vec{p}), M(\omega_{u,n}^0, \vec{p})) & & \\ & \mathrm{diag}_c(M(\omega_{d,n}^-, \vec{p}), M(\omega_{d,n}^+, \vec{p}), M(\omega_{d,n}^0, \vec{p})) & \\ & & \mathrm{diag}_c(M(\omega_{s,n}^-, \vec{p}), M(\omega_{s,n}^+, \vec{p}), M(\omega_{s,n}^0, \vec{p})) \end{pmatrix} \tag{5.3.14a}$$

with

$$\omega_{f,n}^\pm = \omega_n - \mathrm{i}\mu_f \pm A_4^3/2 - A_4^8/(2\sqrt{3}), \omega_{f,n}^0 = \omega_n - \mathrm{i}\mu_f + A_4^8/\sqrt{3}. \tag{5.3.14b}$$

In contrast to the two-flavor model, different quark chemical potentials $\hat{\mu} = \mathrm{diag}_f(\mu_u, \mu_d, \mu_s)$ have now been introduced for the three quark flavors. The trace can be further simplified leading to

$$\Omega = -2T \sum_{f \in \{u,d,s\}} \sum_{i=0,\pm} \sum_{n \in \mathbb{Z}} \int \frac{\mathrm{d}^3 p}{(2\pi)^3} \, \mathrm{Re} \left\{ \ln \left[{\omega_{f,n}^i}^2 + \vec{p}^2 + M^2(\omega_{f,n}^i, \vec{p}) \right] \right\}$$
$$- \frac{1}{2} \left\{ \sum_{f \in \{u,d,s\}} \left(\bar{\sigma}_f \bar{S}_f + \frac{G}{2} \bar{S}_f \bar{S}_f \right) + \frac{H}{2} \bar{S}_u \bar{S}_d \bar{S}_s \right\} + \mathcal{U}(\Phi, \Phi^*, T). \tag{5.3.12'}$$

This is the thermodynamic potential of the nonlocal three-flavor PNJL model in mean-field approximation. The auxiliary scalar fields $\bar{S}_f$ are determined by the SPA conditions, Eq. (3.3.31a).

Gap Equations in Mean-Field Approximation

Once the thermodynamic potential Ω is calculated, the fields $\sigma_u = \sigma_d, \sigma_s$ and A_4^3, A_4^8 are determined by requiring thermodynamic potential to be stationary. The necessary conditions are given by the gap equations

$$\frac{\partial \Omega}{\partial \bar{\sigma}_u} = \frac{\partial \Omega}{\partial \bar{\sigma}_s} = \frac{\partial \Omega}{\partial A_4^3} = \frac{\partial \Omega}{\partial A_4^8} = 0, \tag{5.3.15}$$

together with the stationary phase approximation equations (3.3.31a). First, we limit ourself to the zero-density ($\mu = 0$) case. We then have $A_4^8 = 0$ (cf. Ref. [36]) as discussed in the two-flavor case.

Fig. 5.11 shows the results for the temperature dependence of the chiral up- and strange-quark condensate and of the Polyakov loop using the parameters given in Table 3.2 (scenario I)

---

[16]The factor $\frac{1}{2}$ results again from the doubling of the degrees of freedom in Nambu-Gor'kov space.

and Table 5.1. This figure illustrates once more, as already discussed in the two-flavor case, the entanglement of chiral dynamics and Polyakov-loop degrees of freedom at zero chemical potential. In the absence of a coupling between quark quasiparticles and Polyakov loop the chiral transition (for $N_f = 3$ flavors) and the first-order deconfinement transition (of pure gauge QCD) appear at very different critical temperatures ($T_{\text{chiral}} \approx 110\,\text{MeV}$ for the chiral transition and $T_0 \approx 270\,\text{MeV}$ for deconfinement). The presence of quarks breaks the $Z(3)$ symmetry explicitly and turns the first-order deconfinement phase transition into a continuous crossover. The quark coupling to the Polyakov loop moves this transition to lower temperature. At the same time the chiral transition (with explicit symmetry breaking by nonzero quark mass) turns into a crossover at an upward-shifted temperature, just so that both transitions nearly coincide at a common temperature $T_c \approx 200\,\text{MeV}$.

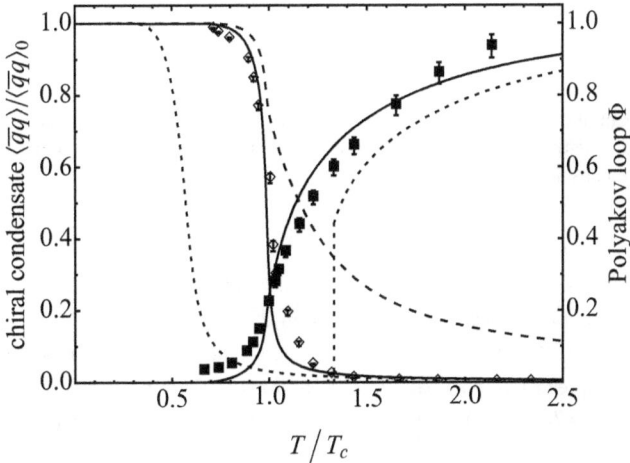

FIGURE 5.11: Results of nonlocal PNJL calculations of the chiral and deconfinement transition pattern using the parameter set of scenario I. Left solid curve: temperature dependence of the chiral condensate $\langle \bar{u}u \rangle = \langle \bar{d}d \rangle$. The strange-quark condensate $\langle \bar{s}s \rangle$ is shown as the dashed curve. Right solid curve: temperature dependence of the Polyakov loop Φ. Left dotted curve: light-quark condensate without coupling to Polyakov loop. Right dotted curve: Polyakov loop in the absence of quarks (pure gauge QCD). Also shown are lattice results for the chiral condensate and the Polyakov loop from the "hotQCD" collaboration [75]. The temperature is given in units of the transition temperature $T_c = 200\,\text{MeV}$.

This symmetry breaking pattern seems also to be realized in recent lattice-QCD results using staggered fermions [75] where a common chiral and deconfinement transition temperature $T_c = (196 \pm 3)\,\text{MeV}$ is observed. Fig. 5.11 shows these lattice data for orientation. For a more detailed discussion, see Sect. 4.3.2. In the latest work of the collaboration [76], domain-wall fermions are used instead of staggered fermions, leading to $T_c = 171(10)(17)\,\text{MeV}$, a value which is consistent with both $T_c = 196\,\text{MeV}$ and the alternative lattice computations of Ref. [77] that find a lower

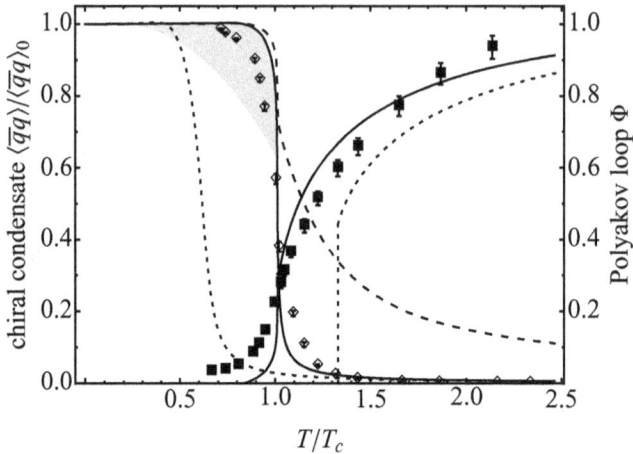

FIGURE 5.12: Chiral condensate and Polyakov loop like in Fig. 5.11, but for parameters of scenario II. The (lower border of the) gray band shows the modifications of the chiral condensate once pionic contributions are considered.

chiral transition temperature, $T_c \simeq 150\,\mathrm{MeV}$.

Compared to the $N_f = 2$ flavor case we observe only minor changes at this point, in particular the transition temperature decreases slightly from $T_c^{2\,\mathrm{f}} \simeq 207\,\mathrm{MeV}$ to $T_c^{3\,\mathrm{f}} \simeq 200\,\mathrm{MeV}$. The transition temperature T_c can be decreased further if the response from quark effects is included in the Polyakov-loop effective potential, leading to a lower T_0 in its parametrization. According to Ref. [126] one has $T_0 = 190\,\mathrm{MeV}$ for $2 + 1$ active quark flavors instead of $T_0 = 270\,\mathrm{MeV}$ for the pure gluon case.

Parameter Dependence

The chiral and deconfinement transition pattern shown in Fig. 5.11 changes only marginally when scenario I (with parameter set listed in Table 3.2) is replaced by scenario II with a coupling G that is about 15 % larger, see Fig. 5.12. It is instructive also to examine the dependence on other parameters such as the current quark mass m_u. The impact of a variation of the pion mass is shown in Figs. 5.13 and 5.14. In the chiral limit, $m_u = m_d \to 0$, the chiral condensate displays a second-order phase transition as expected. Explicit chiral symmetry breaking with $m_{u,d} \neq 0$ turns this into a crossover transition as evident from Fig. 5.11 for $m_{u,d} = 3\,\mathrm{MeV}$. Increasing the quark mass to $m_{u,d} = 10\,\mathrm{MeV}$ makes the crossover softer at $T > T_c$ while leaving the condensate unaltered at temperatures below T_c. The reason is that, below the transition temperature, the dynamical quark mass $M(p)$ entering the chiral condensate in the finite-temperature generalization of Eq. (2.4.9) is dominated by the large scalar field $\bar{\sigma}_u \simeq 0.4\,\mathrm{GeV}$. Changes of the light-quark mass $m_{u,d}$ within a 10 MeV range are not important compared to that scale,

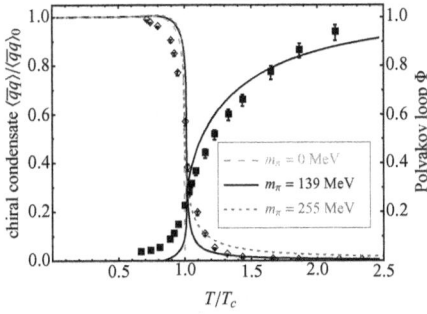

FIGURE 5.13: Pion mass (i. e., current quark mass) dependence of the light chiral condensate as a function of temperature. Parameters of set II were chosen. It is apparent, that higher pion masses influence the behavior of $\langle \bar{q}q \rangle$ only above T_c.

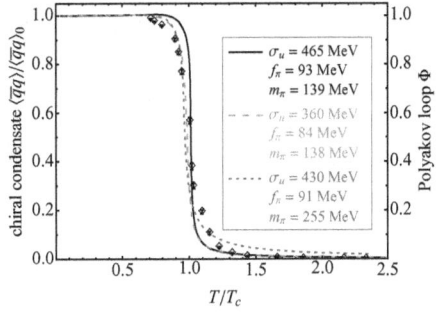

FIGURE 5.14: Dependence of the chiral condensate on the four-fermion coupling strength G. Results for parameter sets I and II are shown for the physical pion mass. In addition, the chiral condensate for set I using a pion mass of $m_\pi \simeq 250\,\text{MeV}$ was calculated and compared to lattice data from Ref. [96] with $m_\pi \simeq 220\,\text{MeV}$.

whereas they become more prominent above T_c where $\bar{\sigma}_u$ drops rapidly. The softening of the strange-quark condensate $\langle \bar{s}s \rangle$ above T_c, as seen in Fig. 5.11, is much more pronounced, given the larger s-quark mass $m_s \simeq 70\,\text{MeV}$.

Corrections to the behavior of the chiral condensate and the pressure below T_c come primarily from thermal pions (and kaons) as will be discussed in the following subsection.

Mesonic Corrections

So far the calculations have been performed in the mean-field approximation in which the pressure $P = -\Omega$ is determined by the quarks moving as quasiparticles in the background provided by the expectation values of the sigma fields, $\bar{\sigma}_u = \bar{\sigma}_d$ and $\bar{\sigma}_s$, and of the Polyakov loop Φ. In order to get a realistic description of the hadronic phase (at temperatures $T \lesssim T_c$), it is important to include mesonic quark-antiquark excitations. The hadronic phase in the absence of baryons is dominated by (the light) pseudoscalar mesons (pions and kaons). The quark-antiquark continuum is suppressed by confinement.

However, as pointed out when discussing the two-flavor case, mesons described within the standard (local) PNJL model can still undergo unphysical decays into the quark-antiquark continuum even below T_c. In the nonlocal PNJL model, such unphysical decays do not appear by virtue of the momentum-dependent dynamical quark mass[17]. This means that the pressure in the $N_f = 3$ case below T_c is now basically generated by the pion and kaon poles of the corresponding one-loop $q\bar{q}$ Green functions.

To include the mesonic contributions to the pressure in our nonlocal PNJL model we can basically use the two-flavor formulation of Sect. 5.3.2. One has to calculate $G_{\text{PS},\text{S}}(\nu_m, \vec{p})$, Eq. (3.3.35),

---

[17]With the possible exception of the η' meson which will not be considered in this section.

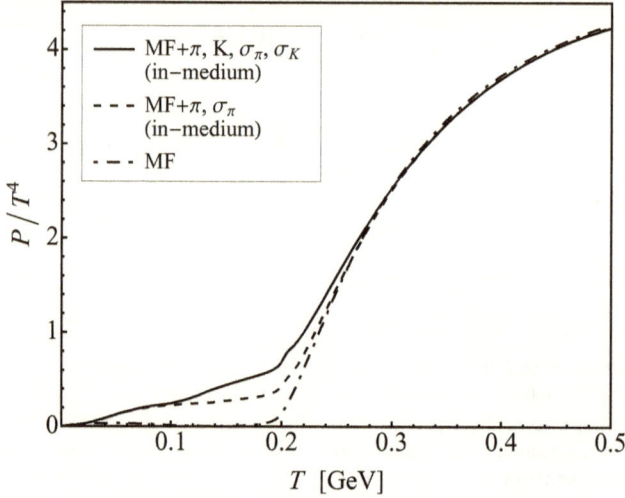

FIGURE 5.15: Pressure (in units of T^4) calculated in the nonlocal PNJL model as a function of temperature on an absolute temperature scale. Solid curve: full calculation (i. e., mean-field result plus mesonic corrections). Dashed-dotted curve: mean-field result (no mesonic corrections). Dashed curve: mean field plus pionic and corresponding scalar modes.

and in particular the quark loop contribution to the pseudoscalar (PS) and scalar (S) mesonic self-energies $\Pi_{\mathrm{PS,S}}(\nu_m, \vec{p})$ (where $\nu_m = 2\pi m T, m \in \mathbb{Z}$ is the bosonic Matsubara frequency and $\vec{p}$ is the momentum of the incoming meson), given in Eq. (3.3.36), at finite temperature. In analogy to Sect. 5.3.2 one has to evaluate the quark loop expression

$$
\Pi_{ij}^{\pm}(\nu_m, \vec{p}) = 8T \sum_{\ell=0,\pm} \sum_{n\in\mathbb{Z}} \int \frac{\mathrm{d}^3 k}{(2\pi)^3}\, \mathcal{C}(\omega_n^\ell, \vec{k}+\vec{p})^2 \times
$$
$$
\times \frac{\left(\omega_n^\ell+\frac{\nu_m}{2}\right)\left(\omega_n^\ell-\frac{\nu_m}{2}\right) + \left(\vec{k}+\frac{\vec{p}}{2}\right)\left(\vec{k}-\frac{\vec{p}}{2}\right) \pm M_i\left(\omega_n^\ell+\frac{\nu_m}{2}, \vec{k}+\frac{\vec{p}}{2}\right) M_j\left(\omega_n^\ell-\frac{\nu_m}{2}, \vec{k}-\frac{\vec{p}}{2}\right)}{\left[\left(\omega_n^\ell+\frac{\nu_m}{2}\right)^2+\left(\vec{k}+\frac{\vec{p}}{2}\right)^2+M_i^2\left(\omega_n^\ell+\frac{\nu_m}{2}, \vec{k}+\frac{\vec{p}}{2}\right)\right]\left[\left(\omega_n^\ell-\frac{\nu_m}{2}\right)^2+\left(\vec{k}-\frac{\vec{p}}{2}\right)^2+M_j^2\left(\omega_n^\ell-\frac{\nu_m}{2}, \vec{k}-\frac{\vec{p}}{2}\right)\right]},
$$
$$
(5.3.16)
$$

with $\omega_n^\pm = \omega_n \pm A_4^3, \omega^0 = \omega_n$ and $M_q(\omega_n, \vec{p}) = m_q + \mathcal{C}(\omega_n, \vec{p})\,\bar{\sigma}_q$. From this we obtain the pressure in random phase approximation (RPA)

$$
P_{\mathrm{meson}}(T) = -T \sum_{M=\mathrm{PS,S}} \frac{d_M}{2} \sum_{m\in\mathbb{Z}} \int \frac{\mathrm{d}^3 p}{(2\pi)^3} \ln\left[G_M(\nu_m, \vec{p})\right], \qquad (5.3.17)
$$

where $G_M(\nu_m, \vec{p})$ are the inverse meson propagators (3.3.38) in the Matsubara formalism and d_M is the mesonic degeneracy factor ($d_M = 3$ for pionic and $d_M = 4$ for kaonic modes).

Results for the pressure in the presence of pion, kaon and scalar modes are presented in Fig. 5.15.[18] Together with the full result (solid line) we show the mean-field result (MF, with

---

[18]The plot shown in Fig. 5.15 uses the parameters of scenario I. The difference between pressure curves

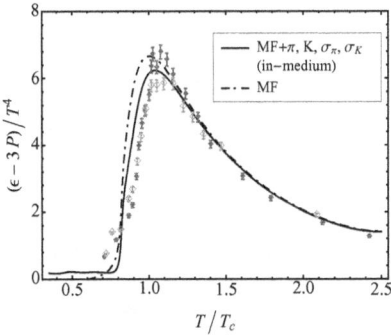

FIGURE 5.16: Trace anomaly $(\epsilon - 3P)/T^4$. Dot-dashed curve: mean-field results. Solid curve: full calculation with inclusion of pions, kaons and scalar modes (results for scenario I). For comparison, three-flavor lattice results for $N_\tau = 8$ from Ref. [96] are shown; the blue points were calculated using the p4 action while the orange points were obtained with the asqtad action.

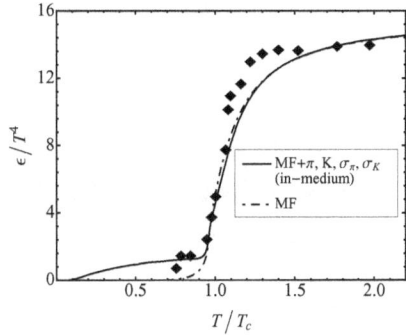

FIGURE 5.17: Energy density ϵ/T^4 for the mean-field and RPA case. (Legend as in Fig. 5.16.) Lattice data ($N_\tau = 8$, p4 action) are taken from Ref. [96].

the pressure determined by quark quasiparticles only) and the mean-field result plus pion and corresponding scalar contributions. It is evident that at low temperatures the mean-field contribution from the quarks is suppressed and the pressure is that of a free meson gas. Near the transition temperature the scalar mesonic modes give a small additional contribution. Finally, above temperatures $T > 1.5 T_c$ the mesonic contributions die out and the quark-gluon mean fields dominate the pressure.

The RPA treatment of mesonic contributions to the pressure allows one, in addition, to compute their corrections to the chiral condensates. Starting from the definition of the chiral condensate,

$$\langle \bar{q}q \rangle = \frac{\int \mathscr{D}A\mathscr{D}\bar{\psi}\mathscr{D}\psi \, \bar{q}q \, e^{-\mathcal{S}_E}}{\int \mathscr{D}A\mathscr{D}\bar{\psi}\mathscr{D}\psi \, e^{-\mathcal{S}_E}} = \frac{\partial \Omega}{\partial m_q}, \qquad (5.3.18)$$

the pionic corrections to $\langle \bar{u}u \rangle$ are computed by differentiating the pion pressure of Eq. (5.3.17) with respect to the up-quark current quark mass:

$$\delta_\pi \langle \bar{u}u \rangle = -\frac{\partial P_\pi}{\partial m_u}. \qquad (5.3.19)$$

From Fig. 5.12 we see, as expected, that the modification of the chiral condensate owing to pions is very similar to the results from chiral perturbation theory (cf., e.g., Refs. [137–139]). At temperatures below T_c pions tend to soften the condensate and make the chiral transition smoother in the range $0.5 T_c < T < T_c$ (see, also, Fig. 7.1).

To conclude this subsection we calculate the energy density, ϵ and the trace anomaly (5.3.11),

calculated from parameter sets I and II turns out to be negligibly small.

$(\epsilon - 3P)/T^4$, for the $N_f = 3$ flavor case, see Figs. 5.16 and 5.17. In contrast to the $N_f = 2$ flavor case it is now sensible to compare our model results to lattice calculations for the $N_f = 2+1$ flavor case from Ref. [75]. Figs. 5.16 and 5.17 show the results that follow from Fig. 5.15 both for the mean-field case and with the additional inclusion of mesonic (pionic and kaonic) contributions.

5.4 Three-Flavor Nonlocal PNJL Model at Finite Density and QCD Phase Diagram

In this section the nonlocal PNJL approach for three flavors is extended to nonzero chemical potentials and the finite-density description of strongly interacting matter. The gap equations (5.3.15) are solved maintaining the chemical potential matrix $\hat{\mu}$ in the effective fermion propagator, Eq. (5.3.13). We restrict ourself to isospin symmetric systems with no strange valence quarks ($\mu_s = 0$) and set $\hat{\mu} = \mathrm{diag}_f(\mu_u, \mu_d, 0)$, $\mu_u = \mu_d$. Furthermore diquark condensates will not be considered.

The introduction of a chemical potential is again accomplished using the prescriptions of the Matsubara formalism (see Sect. 5.1 and Eq. (5.1.9)): the Matsubara frequencies are shifted as $\omega_{f,n} \to \omega_{f,n} - i\mu_f$ in the particle sector (i.e., in the upper-left submatrix) of the Nambu-Gor'kov propagator (5.3.13) and $\omega_{f,n} \to \omega_{f,n} + i\mu_f$ in the corresponding antiparticle sector (i.e., the lower-right submatrix). The thermodynamical potential can then be calculated as in the $N_f = 2$ flavor case.

5.4.1 QCD Phase Diagram

Consider first the T and μ_u dependence of the scalar field $\bar{\sigma}_u$ that acts as a chiral order parameter, deduced from the condition $\frac{\partial \Omega(T,\mu)}{\partial \bar{\sigma}} = 0$. The results are shown in Figs. 5.18 and 5.19 for scenarios I and II, respectively. The profile of $\bar{\sigma}_u$ displays once again the chiral crossover transition at $\mu_u = 0$. It turns into a first-order phase transition at a critical (end) point (CEP) (located at $T_{\mathrm{CEP}} \approx 170\,\mathrm{MeV}$ and $\mu_{\mathrm{CEP}} \approx 180\,\mathrm{MeV}$ for scenario I and at $T_{\mathrm{CEP}} \approx 195\,\mathrm{MeV}$ and $\mu_{\mathrm{CEP}} \approx 110\,\mathrm{MeV}$ for scenario II). The result is qualitatively similar to the the two-flavor case. Quantitatively, however, the location of the critical point is varying significantly depending on the choice of the parameters (see Fig. 4.7). We will discuss this issue in more detail at the end of this chapter.

The projection of Figs. 5.18 and 5.19 onto the T-μ_u plane gives the phase diagram of the nonlocal three-flavor PNJL model, Fig. 5.20. At low μ this phase diagram shows the chiral and deconfinement crossover transitions in close contact, as already discussed. The deconfinement transition is displayed here as a band bounded by dashed lines where the upper and lower bounds are given by values $\Phi = 0.5$ and $\Phi = 0.3$ of the Polyakov loop, respectively, reflecting the relatively soft crossover of this transition (see also Fig. 5.11). At larger values of the chemical potential, beyond the critical point, a separation between the chiral and deconfinement transition takes place.

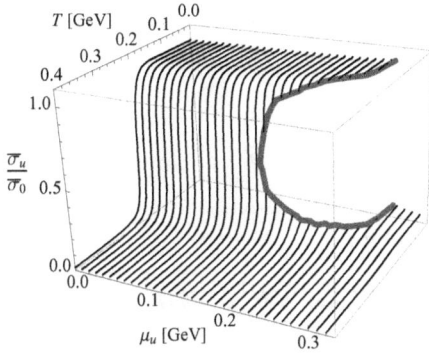

FIGURE 5.18: Chiral order parameter $\bar{\sigma}_u$ for scenario I, normalized to its value $\bar{\sigma}_0$ at $T = \mu = 0$, as a function of temperature and up-quark chemical potential (note $\mu_u = \mu_d, \mu_s = 0$). The blue line shows the border for the first-order transition.

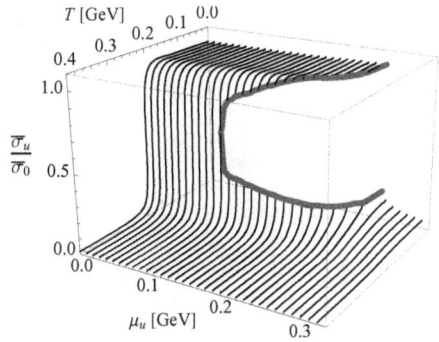

FIGURE 5.19: Same as in Fig. 5.18, for parameters of scenario II. The major difference between Figs. 5.18 and 5.19 is the location of the critical point: $(\mu_{CEP}, T_{CEP})^{(I)} = (180\,\text{MeV}, 170\,\text{MeV})$ compared to $(\mu_{CEP}, T_{CEP})^{(II)} = (115\,\text{MeV}, 195\,\text{MeV})$.

The area between the (first-order) chiral phase transition and the deconfinement crossover has recently been interpreted in terms of a "quarkyonic" phase [81, 140]. From Fig. 5.20 it appears that the chiral first-order transition boundary meets the μ axis at $T = 0$ for values of the baryon chemical potential as small as $\mu_B = 3\mu < 0.9\,\text{GeV}$. This is the domain of nuclear matter that is known to be a Fermi liquid of nucleons. The PNJL model works instead with quarks as quasiparticles, the "wrong" degrees of freedom in this low-temperature phase at moderate baryon densities. Nuclear matter is not covered by PNJL-type models and, consequently, such models cannot be considered realistic at low temperature and baryon chemical potentials $\mu_B = 3\mu \lesssim M_N$ where M_N is the nucleon mass. Chiral effective field theory with baryons is instead the appropriate framework to deal with this part of the phase diagram.

The deconfinement transition band at large μ has been calculated using the Polyakov-loop effective potential (5.2.20). This effective potential does not include higher-order effects due to the presence of quarks at nonzero chemical potential. Such additional μ-dependent effects are expected to move the crossover boundary to lower temperatures as μ increases [126].

Fig. 5.20 demonstrates that the position of the critical point is sensitive to small changes of the four-fermion coupling G. The relatively small increase of this coupling strength between scenarios I and II, keeping hadronic vacuum properties at $T = \mu = 0$ almost unchanged, results nevertheless in a significant shift of the critical point in the T-μ plane.

From Fig. 5.21 it is evident that the phase structure is sensitive to variations of the 't Hooft interaction coupling strength H. An increase of H by only 5 % turns the chiral transition even at $\mu_u = 0$ into a first-order transition. This behavior might be expected considering the Columbia plot (Ref. [68]). On the other hand, a decrease of H, corresponding to a reduced η' mass in the thermal medium, moves the end point to higher chemical potentials and lower temperatures.

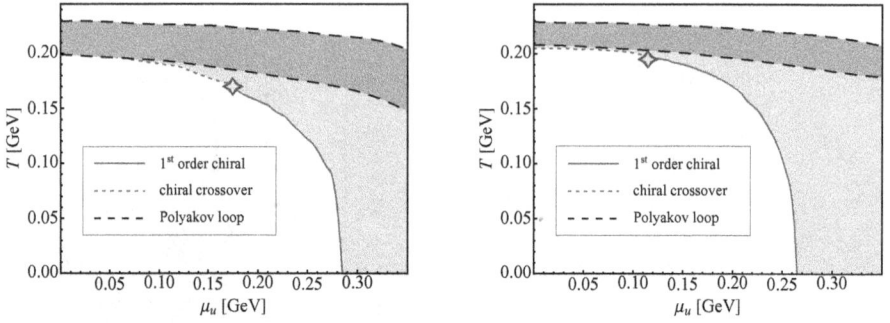

FIGURE 5.20: Phase diagram calculated within the nonlocal PNJL model using the parameters of scenario I (left picture) and scenario II (right picture). The solid blue line shows the first-order chiral transition (the star denotes the critical point). The dashed blue line marks the (chiral) crossover transition while the dashed black lines correspond to the deconfinement transition (the lower and upper lines correspond to $\Phi = 0.3$ and $\Phi = 0.5$, respectively).

The sensitivity to the axial anomaly observed here in the nonlocal PNJL model is, however, less pronounced than that in the local model Ref. [127]. We do not observe that the end point is removed altogether from the phase diagram as quickly as in the local PNJL model.

5.4.2 Pressure at Finite Density

Finally, using Eq. (5.3.12′) with finite chemical potentials included we calculate the pressure $P = -\Omega$ at finite density. In Fig. 5.22 the (normalized) pressure difference

$$\Delta P(T, \mu_u) := P(T, \mu_u) - P(T, \mu_u = 0) \qquad (5.4.1)$$

is shown for selected values of $\mu_u = \mu_d$ and compared to (two-flavor) lattice data from Ref. [141]. Fig. 5.23 displays the full result in the T-μ_u plane. Both figures have been obtained using the parameter set of scenario I.

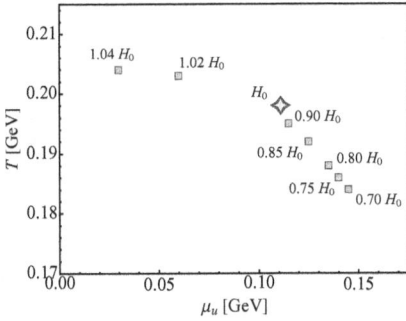

FIGURE 5.21: Location of the critical point depending on several values of the 't Hooft coupling strength H in units of the coupling strength H_0 of scenario II.

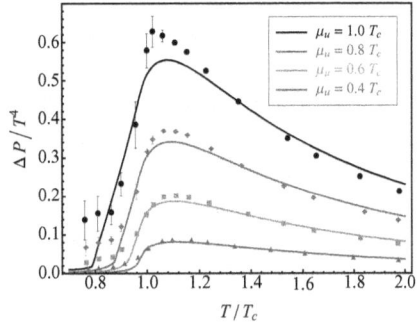

FIGURE 5.22: Pressure difference at finite chemical up-quark potential, $\Delta P(T, \mu_u) = P(T, \mu_u) - P(T, 0)$ (parameters of scenario I) for selected values of μ_u compared to lattice data from Ref. [141].

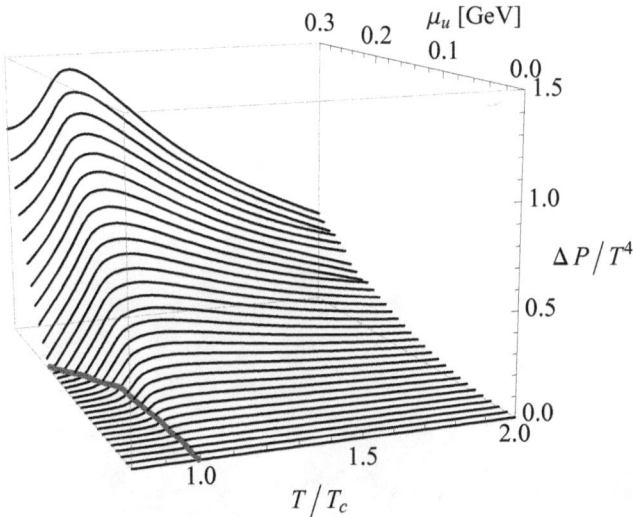

FIGURE 5.23: Excess pressure $\Delta P(T, \mu_u) = P(T, \mu_u) - P(T, 0)$ at finite chemical up-quark potential calculated in the nonlocal PNJL model in units of T^4 for scenario I. The solid blue line indicates the chiral crossover and first-order transition line.

5.4.3 Discussion of the Three-Flavor PNJL Model

Although the nonlocal PNJL model has undoubtedly many virtues, Figs. 5.18, 5.19 and 5.20 should be taken only as qualitative. The location of the critical point is sensitive to the coupling strengths G and H, and to the input current quark mass [37]. The almost constant behavior of $\bar{\sigma}_u(T = 0, \mu_u)$ with increasing quark chemical potential is unrealistic in the absence of explicit baryon (nucleon) degrees of freedom including their interactions.

What is actually required as a starting point for extensions to nonzero chemical potential is a realistic equation of state at finite baryon density, incorporating the known properties of equilibrium and compressed nuclear matter. In such a framework [142], the density dependence of the chiral condensate $\langle \bar{u}u \rangle$ (or of the scalar field $\bar{\sigma}_u$) is well-known to be quite different from the profile shown in Figs. 5.18, 5.19. The magnitude of $\langle \bar{u}u \rangle$ decreases linearly with density ρ [143, 144], with a slope controlled by the pion-nucleon sigma term, and then stabilizes at densities above normal nuclear matter through a combination of two- and three-body correlations and Pauli blocking effects. The transition towards chiral symmetry restoration at $T = 0$ is shifted beyond at least twice the density of normal nuclear matter.

Irrespective of these comments, the nonlocal PNJL approach is obviously instructive in modeling the chiral and deconfinement thermodynamics at $\mu = 0$. Dealing with finite baryon density requires ultimately yet another synthesis, namely a matching of PNJL above the chiral transition and in-medium chiral effective field theory with baryons below that transition.

6 QCD Foundations of the Nonlocal PNJL Model

This final part returns to the question how the nonlocal PNJL model can actually be derived from QCD. We follow here the recent publication of K.-I. Kondo, Ref. [145], extending it to the SU(3) color gauge group. In the last few years, remarkable progress has been made in reformulating QCD [145–152] in terms of the Cho-Faddeev-Niemi-Shabanov (CFNS) decomposition of the Yang-Mills field, proposed first in 1980 by Cho [153, 154] and recently readdressed by Faddeev and Niemi [155, 156], and by Shabanov [157, 158]. This CFNS separation is the basis for the derivation of a nonlocal NJL-type four-point interaction. At the same time it generates an entanglement of the quark sector and the Polyakov-loop variables, A_4^3, A_4^8, exactly as in the nonlocal PNJL model. The nonlocal model considered in this work can, therefore, be derived consistently from full QCD.

6.1 Cho-Faddeev-Niemi-Shabanov Decomposition

Consider first only the Yang-Mills (gauge) part of $\mathscr{L}_{\mathrm{QCD}}$. The Cho-Faddeev-Niemi-Shabanov (CFNS) decomposition is then based on a gauge-independent separation of the non-Abelian gauge field A_μ, Eq. (2.2.4), in terms of a so-called restricted potential, $V_\mu(x)$, of the maximal Abelian subgroup H of the gauge group G (the *maximal torus* H), and the valence potential $X_\mu(x)$ which transforms as a gauge-covariant vector field. The original SU(3) gauge field $A_\mu(x)$ is separated as

$$A_\mu(x) = V_\mu(x) + X_\mu(x), \qquad (6.1.1)$$

such that $V_\mu(x)$ transforms under a SU(3) gauge transformation $U(x)$ identically to the original gauge field $A_\mu(x)$, while $X_\mu(x)$ transforms as an adjoint matter field:

$$V_\mu(x) \to V_{U,\mu}(x) = U(x)\left[V_\mu(x) + \frac{\mathrm{i}}{g}\partial_\mu\right]U^{-1}(x) \qquad (6.1.2a)$$

$$X_\mu(x) \to X_{U,\mu}(x) = U(x)X_\mu(x)U^{-1}(x). \qquad (6.1.2b)$$

The CFNS decomposition is made by introducing a unit color field $\hat{n}(x)$ that is given as a functional of the Yang-Mills gauge field $A_\mu(x)$. Since the decomposition (6.1.1) is required to be gauge independent we demand the field $\hat{n}(x)$ to transform according to the adjoint representation,

$$\hat{n}(x) \to \hat{n}_U = U(x)\hat{n}(x)U^{-1}(x). \qquad (6.1.3)$$

The color field $\hat{n}(x)$ plays a crucial role in the CFNS decomposition. We derive now its general form.

6.1.1 Construction of the Color Vector Field

The *color field* $\hat{n}(x)$ specifies a direction in the color space at each space-time point $x \in \mathbb{R}^4$ in four-dimensional space-time. It is introduced as a unit vector normalized as $\hat{n}(x) \cdot \hat{n}(x) = 1$,[1] and taking real values in the Lie algebra $\mathcal{L}(\mathrm{SU}(3)) = \mathrm{su}(3)$ of the Lie group $\mathrm{SU}(3)$:

$$\hat{n}(x) = n_a(x)t_a, \qquad n_a(x) \in \mathbb{R} \quad (a \in \{1, \dots, 8\}). \tag{6.1.4}$$

Therefore, $\hat{n}(x)$ is a traceless and Hermitian matrix by the properties of the generators $t_a = \frac{\lambda_a}{2}$ of $\mathrm{SU}(3)$ (see Appendix A.2). It follows that $\hat{n}(x)$ can be diagonalized into the matrix D by using a suitable unitary matrix $U(x) \in \mathrm{SU}(3)$:

$$U(x)\hat{n}(x)U^{-1}(x) = \mathrm{diag}(D_1(x), D_2(x), D_3(x)) =: D(x), \tag{6.1.5}$$

with real eigenvalues D_1, D_2, D_3.[2] The diagonal matrix $D(x)$ can be expressed as a linear combination of two diagonal generators H_1, H_2 belonging to the Cartan subalgebra[3], H, as

$$D(x) = U(x)\hat{n}(x)U^{-1}(x) = a(x)H_1 + b(x)H_2, \quad \text{with } a^2 + b^2 = 1. \tag{6.1.6}$$

In $\mathrm{su}(3)$ we can choose $H_1 := t_3$ and $H_2 := t_8$. Therefore the field $\hat{n}(x)$ provides a map[4]

$$\hat{n}(x) : \mathbb{R}^4 \to \mathrm{SU}(3)/H = \mathrm{SU}(3)/(\mathrm{U}(1) \times \mathrm{U}(1)). \tag{6.1.7}$$

Before continuing the discussion of the CFNS decomposition we need to determine the number of independent degrees of freedom carried by the color field $\hat{n}(x)$. From Eq. (6.1.6) it appears convenient to introduce two unit vector fields $\hat{n}_3(x) := U^{-1}(x)t_3U(x)$ and $\hat{n}_8(x) := U^{-1}(x)t_8U(x)$. This choice is, however, not necessarily unique from the viewpoint of reformulating the theory. First notice that the relation (A.2) introduces three types of products; in vector representation of two $\mathrm{SU}(3)$ vectors $\vec{a}, \vec{b} \in \mathrm{SU}(3)$ one has the scalar product (as encountered before), $\vec{a} \cdot \vec{b} = a_a b_a$, the antisymmetric f-product, $(\vec{a} \times \vec{b})_c := f_{abc}a_a b_b$, and the symmetric d-product, $(\vec{a} * \vec{b})_c := d_{abc}a_a b_b$, with the structure constants f_{abc} and d_{abc} as given in Appendix A.2. It turns

---

[1]We use here equivalently the vector form, $\hat{n}(x) = (n_a(x))_{a=1}^8 = (n_1(x), \dots, n_8(x))$, and the Lie-algebra form, $\hat{n}(x) = n_a(x)t_a, a \in \{1, \dots, 8\}$, expressed in terms of the $\mathrm{SU}(3)$ generators $t_a = \frac{\lambda_a}{2}$. For the first form, the scalar product of the generic $\mathrm{SU}(3)$ vectors $\vec{a}, \vec{b}$ is given by $\vec{a} \cdot \vec{b} := a_a b_a$. In the second case, we define the scalar product as $\vec{a} \cdot \vec{b} := 2\mathrm{tr}\left(\vec{a}\vec{b}\right)$. From the relations in Appendix A.2 it is clear that both definitions are equivalent.

[2]We assume here that all three eigenvalues D_1, D_2, D_3 are different. For the case of degenerate eigenvalues see Ref. [149].

[3]The Cartan algebra is the largest commutative subalgebra of a Lie algebra.

[4]Note, that the target space of $\hat{n}(x)$, $\mathrm{SU}(3)/H$, is six-dimensional, instead of the expected seven-dimensional seven-sphere S^7. This is because we require $\mathrm{SU}(3)$ to act *transitively* on the manifold M of the target space; i.e., any two elements of M are required to be connected by a group transformation. If we choose the coset space $\mathrm{SU}(3)/H \subset S^7$, then this requirement is fulfilled.

out that $\hat{n}_3(x), \hat{n}_8(x)$ constitute a closed set of variables under all multiplications $\cdot, \times, *$. In particular, one obtains $\hat{n}_8(x) = \sqrt{3}\,\hat{n}_3(x) * \hat{n}_3(x)$, hence, only $\hat{n}_3(x)$ is an independent variable. Keeping this in mind, one can nevertheless write the unit color field $\hat{n}(x)$ in terms of $\hat{n}_3(x)$ and $\hat{n}_8(x)$. That will be used in the following.

6.1.2 Decomposition

We now return to the CFNS decomposition (6.1.1) and determine the fields $V_\mu(x)$ and $X_\mu(x)$ in terms of the color field $\hat{n}(x)$ and the original gauge field $A_\mu(x)$. As mentioned above, it turns out to be useful to express $\hat{n}(x)$ in terms of the unit fields $\hat{n}_3(x)$ and $\hat{n}_8(x)$ with $\hat{n}_3(x) :=$ $U^{-1}(x)t_3U(x)$ and $\hat{n}_8(x) := U^{-1}(x)t_8U(x)$.[5] These unit vectors commute mutually,

$$[\hat{n}_i(x), \hat{n}_j(x)] = 0 \qquad \text{for } i, j \in \{3, 8\}\,,$$

therefore, $\hat{n}_3(x), \hat{n}_8(x)$ define for a fixed $x \in \mathbb{R}^4$ a basis of the Cartan algebra. The aim is to decompose the gauge field $A_\mu(x)$ into an H-commutative part and a remaining part, lying in $SU(3)/H$. This can be done by using the well known fact that any $su(N)$-Lie-algebra-valued function F can be decomposed into the H-commutative part F_H and the remaining part $F_{SU(N)/H}$:

$$F := F_H + F_{SU(N)/H} := \sum_{i=1}^{r} \hat{n}_i \,(\hat{n}_i \cdot F) + \sum_{i=1}^{r} [\hat{n}_i, [\hat{n}_i, F]]\,, \tag{6.1.9}$$

where $r := \operatorname{rank} SU(N) = N - 1$ and the summation is over the Cartan-algebra elements. In the case of $SU(3)$ we have $r = 2$ and the summation in Eq. (6.1.9) extends over $\hat{n}_3$ and $\hat{n}_8$. We impose the two

Defining Conditions for the Fields $V_\mu(x)$ and $X_\mu(x)$

1. $\hat{n}_3(x) = (n_{3,a}(x))_{a=1}^8$ and $\hat{n}_8(x) = (n_{8,a}(x))_{a=1}^8$ are covariant constants in the background field $V_\mu(x)$:

$$0 = D_\mu[V]\hat{n}_i(x) := \partial_\mu \hat{n}_i(x) - ig[V_\mu, \hat{n}_i(x)]$$
$$= \partial_\mu \hat{n}_i(x) + g f_{abc} V_{a,\mu}(x)\hat{n}_{i,b}(x) \quad \text{for } i \in \{3, 8\} \text{ and } a \in \{1, \ldots, 8\}\,. \tag{6.1.10a}$$

2. $X_\mu(x)$ does not have an H-commutative part, i.e., $X_\mu(x)$ is orthogonal to $\hat{n}_3(x)$ and $\hat{n}_8(x)$:

$$\hat{n}_i(x) \cdot X_\mu(x) = \hat{n}_{i,a}(x)X_{a,\mu}(x) = 0 \quad \text{for } i \in \{3, 8\} \text{ and } a \in \{1, \ldots, 8\}\,. \tag{6.1.10b}$$

These conditions are invariant under the transformations Eq. (6.1.2).

---

[5]Note, that these fields are indeed orthogonal unit vectors, since for $i, j \in \{3, 8\}$ we have

$$\hat{n}_i(x) \cdot \hat{n}_j(x) = 2\operatorname{Tr}\left(U^{-1}(x)H_iU(x)U^{-1}(x)H_jU^{-1}\right) = 2\operatorname{Tr}\left(H_iH_j\right) = \delta_{ij}\,. \tag{6.1.8}$$

The gauge covariant potential $X_\mu(x)$ follows immediately from the second defining condition, Eq. (6.1.10b), and the relation Eq. (6.1.9), applied to $X_\mu(x)$:

$$X_\mu(x) = \frac{1}{g}\hat{n}_3(x) \times D_\mu[A]\hat{n}_3(x) + \frac{1}{g}\hat{n}_8(x) \times D_\mu[A]\hat{n}_8(x)\,, \qquad (6.1.11)$$

where $D_\mu[A]\hat{n}_i(x) := \partial_\mu\hat{n}_i(x) - ig[A_\mu, \hat{n}_i(x)]$ for $i \in \{3, 8\}$. Moreover, the $V_\mu(x)$ field expressed in terms of $A_\mu(x)$ and $\hat{n}_i(x)$ reads

$$V_\mu(x) = (A_\mu(x) \cdot \hat{n}_3(x))\,\hat{n}_3(x) + (A_\mu(x) \cdot \hat{n}_8(x))\,\hat{n}_8 - \frac{1}{g}\hat{n}_3(x) \times \partial_\mu\hat{n}_3(x) - \frac{1}{g}\hat{n}_8(x) \times \partial_\mu\hat{n}_8(x)\,.$$
$$(6.1.12)$$

From the explicit expression for $V_\mu(x)$ it is evident that it can be further decomposed in a part $C_\mu(x)$ with

$$C_\mu(x) := (A_\mu(x) \cdot \hat{n}_3(x))\,\hat{n}_3(x) + (A_\mu(x) \cdot \hat{n}_8(x))\,\hat{n}_8 =: c_3(x)\hat{n}_3(x) + c_{8,\mu}(x)\hat{n}_8(x)\,, \qquad (6.1.13)$$

that is H-commutative,

$$c_{i,\mu}(x) := (A_\mu(x) \cdot \hat{n}_i(x))\,, \qquad (6.1.14)$$

and a remaining part $B_\mu(x)$,

$$B_\mu(x) := -\frac{1}{g}\hat{n}_3(x) \times \partial_\mu\hat{n}_3(x) - \frac{1}{g}\hat{n}_8(x) \times \partial_\mu\hat{n}_8(x)\,, \qquad (6.1.15)$$

that is not H-commutative and perpendicular to $\hat{n}_3(x)$ and $\hat{n}_8(x)$.

Thus, the original gauge field has the following CFNS decomposition:

$$A_\mu(x) = V_\mu(x) + X_\mu(x) = C_\mu(x) + B_\mu(x) + X_\mu(x)\,, \qquad (6.1.16)$$

with the *electric connection* $C_\mu(x)$, Eq. (6.1.13), the *magnetic connection* $B_\mu(x)$, Eq. (6.1.15), and the gauge-covariant *valence potential* $X_\mu(x)$, Eq. (6.1.11).

6.1.3 Transformation Properties of $V_\mu(x)$ and $X_\mu(x)$

The decomposition of the gauge field $A_\mu(x)$ into the sum of $V_\mu(x)$ and $X_\mu(x)$ is gauge independent. We are now going to check the transformation behavior of these fields under (local) SU(3) transformations explicitly. It is sufficient to consider infinitesimal transformations $U(x) = \exp(i\vec{\alpha}_a(x)) = \exp(it_a\alpha_a(x)) = 1 + it_a\alpha_a(x) + \mathcal{O}(\alpha^2)$. First of all, it is important to observe from Eqs. (6.1.11) and (6.1.14), that the local transformations $\delta c_{i,\mu}(x)$ and $\delta X_\mu(x)$ are uniquely determined once the transformations $\delta A_\mu(x)$ and $\delta\hat{n}_i(x)$ are specified. From Eq. (2.2.4) and Eq. (6.1.3), we obtain, respectively

$$\delta A_\mu(x) = \frac{1}{g}D_\mu[A]\vec{\alpha} = \frac{1}{g}\left(\partial\vec{\alpha}(x) + gf_{abc}A_{a,\mu}(x)\alpha_b(x)\right) \qquad (6.1.17a)$$

$$\delta\hat{n}_i(x) = \hat{n}_i(x) \times \vec{\alpha}(x) = f_{abc}\hat{n}_{i,a}(x)\alpha_b(x)\,. \qquad (6.1.17b)$$

Consider first the transformation property of the magnetic connection $B_\mu(x)$:

$$\delta B_\mu(x) = \frac{1}{g} D_\mu[B]\vec{\alpha}(x) - \frac{1}{g}\left[(\hat{n}_3(x)\cdot\partial_\mu\vec{\alpha}(x))\hat{n}_3(x) + (\hat{n}_8(x)\cdot\partial_\mu\vec{\alpha}(x))\hat{n}_8(x)\right]. \qquad (6.1.18)$$

This expression is found by using the Jacobi identity (A.1) and the projection property of the $\hat{n}_i(x)$ (see Ref. [150]),

$$\delta_{ab} = \sum_i \left[\hat{n}_{i,a}\hat{n}_{i,b} - f_{acd}\hat{n}_{i,c}f_{deb}\hat{n}_{i,e}\right]. \qquad (6.1.19)$$

Together with

$$\delta C_\mu(x) = \frac{1}{g}\left[(\hat{n}_3(x)\cdot\partial_\mu\vec{\alpha}(x))\hat{n}_3(x) + (\hat{n}_8(x)\cdot\partial_\mu\vec{\alpha}(x))\hat{n}_8(x)\right] + C_\mu(x)\times\vec{\alpha}(x), \qquad (6.1.20)$$

we find:

$$\delta V_\mu(x) = \frac{1}{g}D_\mu[V]\vec{\alpha}(x), \qquad (6.1.21)$$

and confirm indeed that the restricted connection $V_\mu(x)$ by itself describes an SU(3) connection which incorporates the full SU(3) gauge degrees of freedom. The so-called *restricted gauge theory*, described only by $V_\mu(x)$, is, hence, invariant under SU(3) gauge transformations, although being restricted by condition (6.1.10a).

The transformation behavior of $X_\mu(x)$ is then easily obtained by subtracting Eq. (6.1.21) from Eq. (6.1.17a):

$$\delta X_\mu(x) = X_\mu(x)\times\vec{\alpha}(x) = f_{abc}X_{a,\mu}(x)\alpha_b(x). \qquad (6.1.22)$$

Hence $X_\mu(x)$ transforms covariantly under the gauge transformation.[6] This confirms that the CFNS decomposition provides a gauge-independent decomposition of the non-Abelian potential into the restricted part $V_\mu(x)$ and the gauge-covariant part $X_\mu(x)$.

The SU(3)$^{\vec{\alpha}}_{\text{local}}$ symmetry is, however, not the full gauge symmetry of the (extended) CFNS-Yang-Mills theory. This can be understood from the normalization condition of the color fields, $\hat{n}_i(x)^2 = 1$. It follows that transformations belonging to the *stabilizer group* (or *little group*) of $\hat{n}_3(x)$ and $\hat{n}_8(x)$, SU(3)$_{\hat{n}_i} := \{g \in \text{SU}(3)|g\cdot\hat{n}_i = \hat{n}_i\}, i \in \{3,8\}$, leave $\hat{n}_i(x)$ invariant for a given *fixed* $A_\mu(x)$. From Eqs. (6.1.14) and (6.1.11) it is then obvious that the CFNS theory is invariant under these conditions. Noting SU(3)$_{\hat{n}_3,\hat{n}_8} = H = \text{U}(1)\times\text{U}(1)$, the CFNS-Yang-Mills theory is invariant under additional SU(3)/(U(1) × U(1)) gauge transformations. Assigning the gauge parameter $\vec{\theta}$ to this symmetry, the CFNS-Yang-Mills theory then has the local gauge symmetry

$$G^{\vec{\alpha},\vec{\theta}}_{\text{local}} := \text{SU}(3)^{\vec{\alpha}}_{\text{local}} \times [\text{SU}(3)/(\text{U}(1)\times\text{U}(1))]^{\vec{\theta}}_{\text{local}} \qquad (6.1.23)$$

which is larger than the local SU(3)$^{\vec{\alpha}}_{\text{local}}$ symmetry of the original Yang-Mills theory. One has to introduce some gauge-fixing procedure in order to reduce this extended symmetry to the SU(3) gauge symmetry of QCD. This will be described in Sect. 6.1.5.

---

[6]This is not surprising, because the connection space, i.e., the space of all gauge potentials, forms an affine space. This guarantees that one can describe an arbitrary potential by adding a gauge-covariant piece $X_\mu(x)$ to the restricted potential.

6.1.4   Physical Relevance of the CFNS Decomposition and Wilson Loop

To understand the physical meaning of the CFNS decomposition we first express the field strength tensor $G_{\mu\nu}$, Eq. (2.2.2), in terms of the new fields $V_\mu(x)$ and $X_\mu(x)$:

$$G_{\mu\nu} = \frac{i}{g}\left[D_\mu[A], D_\nu[A]\right]$$
$$= \underbrace{(F_{3,\mu\nu} + H_{3,\mu\nu})\hat{n}_3 + (F_{8,\mu\nu} + H_{8,\mu\nu})\hat{n}_8}_{=:\hat{G}_{\mu\nu}} + D_\mu[V]X_\nu - D_\nu[V]X_\mu + gX_\mu \times X_\nu \,, \qquad (6.1.24)$$

The field strength tensor of the restricted theory, $\hat{G}_{\mu\nu}$, is composed of the field strengths $F_{i,\mu\nu} + H_{i,\mu\nu}$ made of the restricted potential,

$$F_{i,\mu\nu} := \partial_\mu c_{i,\nu} - \partial_\nu c_{i,\mu} \qquad (6.1.25a)$$

$$H_{i,\mu\nu} := -\frac{1}{g}\hat{n}_i \cdot (\partial_\mu \hat{n}_i \times \partial_\nu \hat{n}_i) =: \partial_\mu \tilde{c}_{i,\nu} - \partial_\nu \tilde{c}_{i,\mu} \,, \qquad (6.1.25b)$$

where $\tilde{c}_{i,\mu}$ are the *magnetic potentials*.[7] Some comments are in order: first notice, that—not surprisingly—the field strength tensor

$$\hat{G}_{\mu\nu} := \frac{i}{g}\left[D_\mu[V], D_\nu[V]\right] \,, \qquad (6.1.26)$$

defined by the restricted connection is proportional to $\hat{n}_3(x)$ and $\hat{n}_8(x)$. This actually follows directly from the definition of the field strength tensor and from the defining condition (6.1.10a):

$$\hat{G}_{\mu\nu} \times \hat{n}_i = \frac{1}{g}\left[D_\mu[V], D_\nu[V]\right]\hat{n}_i = 0 \,,$$

which is only fulfilled if $G_{\mu\nu}$ is proportional to $\hat{n}_3(x)$ and $\hat{n}_8(x)$. From Eq. (6.1.25) the dual structure of the restricted connection $V_\mu(x)$ becomes evident; this justifies the notations electric and magnetic potentials for $c_{i,\mu}$ and $\tilde{c}_{i,\mu}$, respectively. This dual structure allows one to identify (color) electric and magnetic currents, j_μ and k_μ, by applying Maxwell's equations to the field strength $G_{\mu\nu}$ and its dual form $*G_{\mu\nu} := \frac{1}{2}\varepsilon_{\mu\nu\rho\sigma}G_{\rho\sigma}$:

$$j_{i,\mu}(x) = \partial_\nu\left(\hat{n}_i(x) \cdot G_{\mu\nu}\right) = \partial_\nu F_{i,\mu\nu} \qquad (6.1.27a)$$

$$k_{i,\mu}(x) = \partial_\nu\left(\hat{n}_i(x) \cdot *G_{\mu\nu}\right) = \partial_\nu *H_{i,\mu\nu} \,, \qquad (6.1.27b)$$

where $*H_{i,\mu\nu} = \frac{1}{2}\varepsilon_{\mu\nu\rho\sigma}H_{i,\rho\sigma}$ is the dual magnetic field strength tensor.

---

[7]Note, that the existence of such potentials is not clear at first sight. For the existence, de Rahm's theorem, $\partial_\mu\left(\varepsilon_{\mu\nu\rho\sigma}H_{\rho\sigma}\right) = 0$, has to be fulfilled. We will see below, however, that the divergence of the dual magnetic field strength tensor is determined by the magnetic monopole current which is not identically zero, but contains singularities. In Ref. [159] it was shown, however, that such potentials can be found at least locally in each *section of the fiber bundle of* $(\mathbb{R}^4, \mathrm{SU}(3)/(\mathrm{U}(1) \times \mathrm{U}(1)))$. Since this proof relies on the *fiber-bundle* description of the restricted gauge theory, it is beyond the scope of this work here.

Magnetic Monopoles

Taking the derivatives of the currents, one finds $\partial_\mu j_i^\mu(x) = \partial_\mu k_i^\mu(x) = 0$. It is tempting to expect that the magnetic charges

$$g_{\hat{n}_i} = \int_{S_R^2} k_{i,0}(x)\, \mathrm{d}\vec{\sigma}\,, \tag{6.1.28}$$

are conserved, where the integration is carried out over the two-sphere S_R^2 (see, e. g., Refs. [160, 161]): $\vec{x}^2 = R^2$ with $R \to \infty$. This argument is, however not true here. Although the magnetic charge is indeed conserved, this does not originate in dynamics, but rather follows from the topological structure of the color fields $\hat{n}_3(x)$ and $\hat{n}_8(x)$ in the three-dimensional space.[8] This means, that the $g_{\hat{n}_i}$ do not generate a symmetry, and so Noether's theorem does not apply. This can be seen by noticing that the $g_{\hat{n}_i}$ are completely determined by the color fields $\hat{n}_i(x)$ and their derivatives. They commute with all dynamical variables; thus, they do not generate symmetry transformations. The appearance of a conserved (topological) magnetic charge is owing to nontrivial boundary conditions of the color fields $\hat{n}_i(x)$. We have mentioned at the beginning of this section, that for $(\hat{n}_3(x), \hat{n}_8(x))$ being transitive, their target space is required to be restricted to $\mathrm{SU}(3)/(\mathrm{U}(1) \times \mathrm{U}(1))$. On the other hand, the integral in Eq. (6.1.29) is extended over the two-sphere. Therefore, for $(\hat{n}_3(x), \hat{n}_8(x))$ being singe-valued, as the two-sphere is covered once, each of the $\hat{n}_3(x), \hat{n}_8(x)$ will be covered an integral number of times. This is the result of

$$\pi_2(\mathrm{SU}(3)/(\mathrm{U}(1) \times \mathrm{U}(1))) \simeq \pi_1(\mathrm{U}(1) \times \mathrm{U}(1)) \simeq \mathbb{Z} \times \mathbb{Z}\,,$$

where $\pi_2(G/H), \pi_1(G/H)$ denote the *second* and *first homotopy group* of G/H. A nontrivial second homotopy group of the quotient group of the gauge group G modulo an unbroken subgroup H, $\pi_2(G/H) \neq 0$, is actually a very general condition for the appearance of magnetic monopoles in a gauge theory (see Refs. [162, 163]). From this, one obtains the following magnetic charges:

$$\begin{aligned} g_{\hat{n}_3} &= \int_{S_R^2} k_{3,0}(x)\, \mathrm{d}\vec{\sigma} = \frac{4\pi}{g}\left(m - \frac{1}{2}m'\right) \\ g_{\hat{n}_8} &= \int_{S_R^2} k_{8,0}(x)\, \mathrm{d}\vec{\sigma} = \frac{1}{2}\sqrt{3}\frac{4\pi}{g}m'\,, \end{aligned} \tag{6.1.29}$$

where $m, m' \in \mathbb{N}$.[9]

Monopole Condensation and Confinement

The restricted connection $V_\mu(x)$ describes the potentials of electric and magnetic monopoles. Now we examine the physical meaning of the covariant field $X_\mu(x)$. Writing the gauge part of the QCD Lagrangian, Eq. (2.2.1), in terms of the CFNS variables $\hat{n}_i(x), V_\mu(x), X_\mu(x)$, the

---

[8]We remind the reader of the discussion of instantons in Sect. 3.3.1, and, in particular, of the Chern-Simons current.

[9]As mentioned earlier, $\hat{n}_3(x)$ and $\hat{n}_8(x)$ are not independent; this is now manifest in the magnetic charges, too.

(Euclidean) Yang-Mills Lagrangian becomes

$$\mathscr{L}_{\text{gauge}} = \frac{1}{2} \text{Tr} \left(G_{\mu\nu} G^{\mu\nu} \right)$$
$$= \frac{1}{4} \left[(F_{3,\mu\nu} + H_{3,\mu\nu})^2 + (F_{8,\mu\nu} + H_{8,\mu\nu})^2 \right] + \frac{1}{2} X_{a,\mu} \mathcal{Q}_{\mu\nu}^{ab} X_{b,\nu} + \tag{6.1.30}$$
$$- \frac{1}{2} \left(D_\mu[V] X_\nu - D_\nu[V] X_\mu \right) \cdot g(X_\mu \times X_\nu) + \frac{1}{4} g^2 (X_\mu \times X_\nu)^2 ,$$

where we have defined

$$\mathcal{Q}_{\mu\nu}^{ab}[V] := -(D_\rho[V] D_\rho[V])^{ab} \delta_{\mu\nu} + 2g f_{abc} \hat{G}_{\mu\nu}^c , \tag{6.1.31}$$

with

$$-(D_\rho[V] D_\rho[V])^{ab} = -\delta^{ab} \partial_\rho^2 + g^2 \left[\left(V_\rho^c \right)^2 \delta^{ab} - V_\rho^a V_\rho^b \right] - g f_{abc} \partial_\rho V_\rho^c - 2g f_{abc} V_\rho^c \partial_\rho . \tag{6.1.32}$$

The expressions above follow by noting $\hat{n}_i(x)^2 = 1$, $\hat{n}_3(x) \cdot \hat{n}_8(x) = 0$ and by using first the defining condition (6.1.10b) and then (6.1.10a):

$$0 = D_\mu[V](\hat{n}_i(x) \cdot X_\nu(x)) = \hat{n}_i(x) \cdot D_\mu[V] X_\nu(x) + D_\mu[V] \hat{n}_i(x) \cdot X_\nu(x) = \hat{n}_i(x) \cdot D_\mu[V] X_\nu(x) .$$

It is evident that Eq. (6.1.30) describes a U(1) × U(1) gauge theory coupled to a charged vector field $X_\mu(x)$. In the literature, $X_\mu(x)$ is, hence, interpreted as the "valence gluon" which plays the role of a colored source of the restricted theory [153, 154]. Moreover, $X_\mu(x)$ represents, apart from the quarks, an independent colored degree of freedom that is subject to confinement. This is a hint why only the restricted connection, Eq. (6.1.12), should play the dominant role for confinement. The Lagrangian (6.1.30) has already been ordered according to powers of the valence-gluon field $X_\mu(x)$. If we restrict ourself to quadratic terms in Eq. (6.1.32) without derivatives, we obtain $\frac{1}{2} g^2 \left[\left(V_\rho^c \right)^2 \delta^{ab} - V_\rho^a V_\rho^b \right] X_\mu^a X_\mu^b$, which can be diagonalized leading to a mass term for the valence gluons. It turns out, then (see Ref. [147]), that these gluons acquire a nonzero mass M_X, if vacuum condensation[10]

$$\langle B_\rho(x) \cdot B_\rho(x) \rangle \neq 0$$

occurs. This mass term is $M_X^2 \sim g^2 \langle B_\rho(x) \cdot B_\rho(x) \rangle$. From lattice calculations for the SU(2) case it turns out to be $M_X \simeq 1.2\,\text{GeV}$ (see Ref. [164]).

Two important observations follow immediately from this result: first, remember that the magnetic connection $B_\mu(x)$ is related to the magnetic monopole potential. Hence, $\langle B_\mu(x) \cdot B_\rho(x) \rangle \neq 0$ is also related to the onset of magnetic monopole condensation which provides a possible mechanism for confinement, as follows [165]: consider first ordinary superconductors which are characterized by condensation of (equal) electric charges, the Cooper pairs. It is well known that magnetic fields cannot penetrate a superconductor because of the Meissner effect.

---

[10]In this discussion, possible condensates of the electric components, $\langle C_\rho(x) \cdot C_\rho(x) \rangle$, are neglected, because they are incompatible with the residual U(1) × U(1) invariance.

This can be understood from the formation of a magnetic flux tube [166] (Abrikosov string) that connects a magnetic monopole and an antimonopole by a linear potential.

Polyakov Loop

The Wilson-loop operator is defined as a path-ordered product of an exponent along a closed loop C. With the non-Abelian Stokes theorem derived by Diakonov and Petrov [167, 168] one obtains the following expression for the Wilson loop $\mathcal{W}_A$: Let G be a non-Abelian group with maximal torus group H. Let $H_i \in H$ $(i \in \{1, \dots, r\}, r = \operatorname{rank} G)$ be the generators of the Cartan subalgebra of the corresponding Lie algebra of G. Finally, let $\vec{m} := (m_1, \dots, m_r)$ denote the *highest weight* of the chosen representation. Then (cf. also Ref. [169]),

$$
\begin{aligned}
\mathcal{W}_A(C) &:= \operatorname{Tr}\left[\mathcal{P}\exp\left(\mathrm{i}\oint_C A_\mu(x)\,\mathrm{d}x^\mu\right)\right] \\
&= \int \mathscr{D}U(\xi)\exp\left\{\mathrm{i}\oint \mathrm{d}\xi\,\operatorname{Tr}\left[m_iH_i\left(U(\xi)A(\xi)U^{-1}(\xi)+\frac{\mathrm{i}}{g}U(\xi)\frac{\mathrm{d}}{\mathrm{d}\xi}U^{-1}(\xi)\right)\right]\right\},
\end{aligned}
\tag{6.1.33}
$$

where $A(\xi) := A_\mu(\xi)\frac{\mathrm{d}x^\mu}{\mathrm{d}\xi}$. Here the functional integration extends over all gauge configurations $U(\xi)$.[11] Formula (6.1.33) is manifestly gauge invariant, as is the Wilson loop itself. This version of the Wilson loop removes the path ordering from $\mathcal{W}_A$ in favor of a functional integration over all gauge transformations $U(\xi)$ along the loop.

We are now going to show that the exponential in Eq. (6.1.33) is already determined by the restricted connection $V_\mu(x)$, Eq. (6.1.12). First notice that from $\partial_\mu\hat{n}_i(x) = [\partial_\mu U^{-1}(x)\,U(x), \hat{n}_i(x)]$ and the defining condition (6.1.10a) one obtains $[V_\mu(x) + \mathrm{i}g^{-1}\partial_\mu U^{-1}(x)\,U(x), \hat{n}_i(x)] = 0$ for all $i \in \{1, \dots, r\}$. From Schur's lemma, this implies, that $V_\mu(x)$ must have the form

$$
V_\mu(x) = \Gamma_{i,\mu}(x)\hat{n}_i(x) + \mathrm{i}g^{-1}U^{-1}(x)\,\partial_\mu U(x)\,.
$$

Comparing this to Eq. (6.1.24) it follows that $\Gamma_{i,\mu}(x) = c_{i,\mu} + \tilde{c}_{i,\mu}$, a consequence of the dual nature of $V_\mu(x)$. Furthermore, the second term can be expanded in terms of the a Cartan decomposition using $\hat{n}_i(x), i \in \{1, \dots, r\}$ and $\hat{n}_{\perp,i}(x), i \in \{1, \dots, \dim G - r\}$ (with $\hat{n}_i(x) \cdot \hat{n}_{\perp,j}(x) = 0$ and $\hat{n}_i(x) \cdot \hat{n}_i(x) = \hat{n}_{\perp,j}(x) \cdot \hat{n}_{\perp,j}(x) = 1$) as a basis of G. Hence, the original connection may be

---

[11]For the Wilson loop (and the Polyakov loop, considered later, as well) the gauge configurations $U(\xi)$ are subject to periodic boundary conditions: assume ξ_1 being one point on the closed contour and let ξ_2 correspond to the same point but after we have performed a loop. Then the integration limits in Eq. (6.1.33) are explicitly given as

$$
\int \mathscr{D}U(\xi)\exp\left(\mathrm{i}\oint \mathrm{d}\xi\dots\right) \equiv \int \mathrm{d}U \int_{U(\xi_1)=U}^{U(\xi_2)=U} \mathscr{D}U(\xi)\exp\left(\mathrm{i}\int_{\xi_1}^{\xi_2}\mathrm{d}\xi\dots\right),
$$

where $\mathrm{d}U$ denotes the invariant Haar measure on G/H. For the Polyakov loop, the boundary conditions are $\xi_1 := \xi(0) = \xi(\beta) =: \xi_2$.

written as

$$A_\mu(x) = (c_{i,\mu}(x) + \tilde{c}_{i,\mu}(x))\hat{n}_i(x) + \frac{2\mathrm{i}}{g}\mathrm{Tr}\left(\hat{n}_a(x)U^{-1}(x)\,\partial_\mu U(x)\right)\hat{n}_a(x)$$
$$+ \frac{2\mathrm{i}}{g}\mathrm{Tr}\left(\hat{n}_{\perp,a}(x)U^{-1}(x)\,\partial_\mu U(x)\right)\hat{n}_{\perp,a}(x) + X_\mu(x)\,, \tag{6.1.34}$$

where we have used the previously introduced convention for the scalar product $\vec{a}\cdot\vec{b} = 2\mathrm{Tr}(\vec{a}\,\vec{b})$. Next, again from the projection property

$$m_i\hat{n}_i(x)\cdot A_\mu(x) = 2\mathrm{Tr}\left[m_i H_i U(x)A_\mu U^{-1}(x)\right]\,,$$

with

$$m_i\hat{n}_i(x)\cdot A_\mu(x) = m_i(c_{i,\mu}(x) + \tilde{c}_{i,\mu}(x)) + \frac{2\mathrm{i}}{g}\mathrm{Tr}\left(m_i\hat{n}_i(x)U^{-1}(x)\frac{\mathrm{d}U}{\mathrm{d}x}\right)$$
$$= m_i(c_{i,\mu}(x) + \tilde{c}_{i,\mu}(x)) + \frac{2\mathrm{i}}{g}\mathrm{Tr}\left(m_i H_i \frac{\mathrm{d}U}{\mathrm{d}x}U^{-1}(x)\right)$$

we finally obtain[12]

$$\mathrm{Tr}\left[m_i H_i\left(U(x)A_\mu(x)U^{-1}(x) + \frac{\mathrm{i}}{g}U(x)\frac{\mathrm{d}}{\mathrm{d}x}U^{-1}(x)\right)\right] = \frac{1}{2}m_i(c_{i,\mu}(x) + \tilde{c}_{i,\mu}(x))\,. \tag{6.1.35}$$

The Polyakov loop can then be calculated from

$$\mathcal{W}_A(C) = \int \mathrm{d}U(\xi)\,\exp\left\{\mathrm{i}\oint \mathrm{d}\xi\left[\frac{1}{2}m_i(c_i(\xi) + \tilde{c}_i(\xi))\right]\right\}\,, \tag{6.1.36}$$

with $c_i(\xi) := c_{i,\mu}\frac{\mathrm{d}x^\mu}{\mathrm{d}\xi}, \tilde{c}_i(\xi) := \tilde{c}_{i,\mu}\frac{\mathrm{d}x^\mu}{\mathrm{d}\xi}$. Thus the Wilson line in Eq. (6.1.33) is given by the diagonal components ("Abelian projection") of the connection $A_\mu(x)$. The off-diagonal gluons, $X_\mu(x)$, appear only in the expectation value of the Wilson loop, $\langle\mathcal{W}_A(C)\rangle$, as a term modifying the Yang-Mills Lagrangian (6.1.30). This concludes the proof of Abelian dominance based on monopole condensation.

6.1.5 Reduction Scheme

Having clarified the physical relevance of the CFNS decomposition, we can now proceed to show how the nonlocal PNJL model derives from QCD. As noted previously (cf. Eq. (6.1.23)), the CFNS-Yang-Mills theory has an enlarged gauge symmetry,

$$G_{\text{local}}^{\vec{\alpha},\vec{\theta}} = \mathrm{SU}(3)_{\text{local}}^{\vec{\alpha}} \times [\mathrm{SU}(3)/(\mathrm{U}(1)\times\mathrm{U}(1))]_{\text{local}}^{\vec{\theta}}\,. \tag{6.1.23}$$

On the other hand, the number of degrees of freedom in the CFNS-Yang-Mills theory is 6 from $\hat{n}(x)$, $4\cdot 2 = 8$ from $c_{3,\mu}(x), c_{8,\mu}(x)$, and $4\cdot 6 = 24$ for $X_\mu(x)$ because of Eq. (6.1.10b). This implies that there are $32 + 6$ degrees of freedom in total, 6 more than in standard quantum chromodynamics. This excess number of degrees of freedom is introduced by the color field $\hat{n}(x)$

---

[12]Use $\frac{\mathrm{d}U^{-1}}{\mathrm{d}\xi} = -U^{-1}\frac{\mathrm{d}U}{\mathrm{d}\xi}U^{-1}$.

that is also responsible for the enlargement of the gauge symmetry. However, as pointed out previously, $\hat{n}(x)$ has only topological character but it is not a dynamical variable: the CFNS gauge Lagrangian (6.1.30) does not generate an equation of motion for $\hat{n}(x)$. In this sense, $\hat{n}(x)$ is a gauge artifact that can be removed by a gauge transformation.[13] This means that the additional degrees of freedom *and* the local $[\mathrm{SU}(3)/(\mathrm{U}(1) \times \mathrm{U}(1))]^{\vec{\theta}}$ symmetry can be removed by imposing a gauge condition on the CFNS-Yang-Mills theory.

The gauge-fixing condition that reduces the gauge symmetry of the CFNS-Yang-Mills Lagrangian to the SU(3) local symmetry of the ordinary QCD Lagrangian proceeds as follows (see Refs. [146, 149]). First, recall from Sect. 6.1.3 that the transformation properties of the decomposed fields $B_\mu(x), C_\mu(x), X_\mu(x)$ are uniquely determined once those for $A_\mu(x)$ and $\hat{n}_i(x)$ are specified. Furthermore, the gauge parameters $\vec{\alpha}(x)$ and $\vec{\theta}(x)$ are independent, since the original Yang-Mills Lagrangian does not depend on the choice of $\vec{\theta}(x)$. Therefore, the infinitesimal version of the enlarged gauge transformation, $\delta_{\vec{\alpha},\vec{\theta}}$, can be obtained by combining the local transformations for $\delta_{\vec{\alpha}}A_\mu(x)$ and $\delta_{\vec{\theta}}\hat{n}_i(x)$ (see Sect. 6.1.3):

$$\delta_{\vec{\alpha}}A_\mu(x) = D_\mu[A]\vec{\alpha}(x) \tag{6.1.37a}$$

$$\delta_{\vec{\theta}}\hat{n}_i(x) = g\hat{n}_i(x) \times \vec{\theta}_\perp(x), \tag{6.1.37b}$$

where $\vec{\theta}_\perp \in \mathcal{L}(\mathrm{SU}(3)/H)$. Comparing this to Eqs. (6.1.17), it is clear that the enlarged symmetry (6.1.23) is broken down to the SU(3) symmetry if the condition $\vec{\alpha}_\perp(x) = \vec{\theta}_\perp(x)$, with $\vec{\alpha}_\perp(x) := \vec{\alpha}(x) - (\vec{\alpha} \cdot \hat{n}(x))\hat{n}(x)$, holds. The appropriate reduction condition is then a version of the *maximal Abelian gauge*, demanding a minimizing condition on the square of $X_\mu(x) \cdot X_\mu(x)$ with respect to variations of $\vec{\alpha}(x)$ and $\vec{\theta}(x)$:

$$\delta_{\vec{\alpha},\vec{\theta}} \int \mathrm{d}^4x \, \frac{1}{2} X_\mu(x) \cdot X_\mu(x) = 0. \tag{6.1.38}$$

Using Eq. (6.1.19) and the Jacobi identity (A.1) it is easy to show that

$$g^2 X_\mu(x) \cdot X_\mu(x) = (D_\mu[A]\hat{n}_i(x)) \cdot (D_\mu[A]\hat{n}_i(x)) .$$

It follows by means of Eqs. (6.1.37) that

$$\delta_{\vec{\alpha},\vec{\theta}} \left\{ \frac{1}{2} \left(D_\mu[A]\hat{n}_i(x) \right)^2 \right\} = g \left(\hat{n}_i \times D_\mu[A]\hat{n}_i(x) \right) \cdot D_\mu[A](\vec{\alpha}_\perp(x) - \vec{\theta}_\perp(x)) .$$

Clearly, $(D_\mu[A]\hat{n}_i(x))^2$ is invariant under the subset $\vec{\alpha}_\perp(x) = \vec{\theta}_\perp(x)$ of the enlarged transformation (6.1.37). Therefore, this gauge-fixing condition leaves exactly the SU(3) gauge symmetry invariant.

Finally, we write the gauge condition (6.1.38) in its differential form. This can be deduced from Eq. (6.1.38) using integration by parts and using the definition (6.1.11) for the covariant

---

[13]In order to be precise, this statement is true at least locally section-wise, as discussed above.

field X_μ:

$$\delta_{\vec{\alpha},\vec{\theta}} \int d^4x \left\{ \frac{1}{2} (D_\mu[A]\hat{n}_i(x))^2 \right\} = - \int d^4x \left(\vec{\alpha}_\perp(x) - \vec{\theta}_\perp(x) \right) \cdot D_\mu[A] g \left(\hat{n}_i \times D_\mu[A]\hat{n}_i(x) \right)$$
$$= - \int d^4x \left(\vec{\alpha}_\perp(x) - \vec{\theta}_\perp(x) \right) \cdot g D_\mu[A] X_\mu(x) .$$

The differential reduction condition, χ, is then expressed as[14]

$$\chi[\{\hat{n}_i\}, C_\mu, X_\mu] := D_\mu[V]X_\mu(x) = 0 . \tag{6.1.39}$$

Indeed, $\chi \in \mathrm{SU}(3)/(\mathrm{U}(1) \times \mathrm{U}(1))$ (one has $\hat{n}_i(x) \cdot \chi(x) = 0$ because of $0 = D_\mu[V](\hat{n}_i(x) \cdot X_\mu(x)) = \hat{n}_i(x) \cdot D_\mu[V]X_\mu(x))$, and hence the number of reduction conditions is $8 - 2 = 6$, as necessary.

The major advantage of the gauge-fixing condition (6.1.39) is clearly that it removes the local $\mathrm{SU}(3)/(\mathrm{U}(1) \times \mathrm{U}(1))$ symmetry from the theory, leaving the remaining $\mathrm{SU}(3)$ gauge symmetry fully intact. A second gauge condition must be chosen in order to remove the unphysical QCD degrees of freedom from the theory (imposing, i. e., the Landau or Lorenz gauge condition). One is really left with ordinary quantum chromodynamics expressed in terms of the new CFNS variables $\hat{n}_i(x), c_{i,\mu}(x), X_\mu(x)$. The requirement of a gauge-fixing condition for $X_\mu(x)$ is not surprising. This feature is familiar from the background-field formalism. Indeed, $V_\mu(x)$ acts here as the background field. The decomposition $A_\mu(x) = V_\mu(x) + X_\mu(x)$ is not unique since one can always add another covariant field $\tilde{X}_\mu(x)$ and find an alternative decomposition. In order to get rid of this arbitrariness one needs to fix $X_\mu(x)$ by imposing a gauge condition, as given by Eq. (6.1.39).

6.2 Derivation of the Nonlocal PNJL Model

We are now ready to express the (Euclidean) QCD action, Eq. (2.5.6), in terms of the CFNS fields $\hat{n}_i(x), c_{i,\mu}(x), X_\mu(x)$ $(i \in \{3, 8\})$. Originally,

$$S_E^{\mathrm{QCD}} = S_q + S_{\mathrm{gauge}}$$
$$S_q = \int d^4x \sum_{f \in \mathrm{flavors}} \bar{\psi}_f \left(-i\partial\!\!\!/ + m_f + g A\!\!\!/ + i\mu_f\gamma_4 \right) \psi_f \tag{2.5.6'}$$
$$S_{\mathrm{gauge}} = \frac{1}{2} \int d^4x \mathrm{Tr}(G_{\mu\nu}G_{\mu\nu}) .$$

After applying the CFNS decomposition $A_\mu(x) = V_\mu(x) + X_\mu(x)$. The matter part S_q becomes:

$$S_q = \int d^4x \left[\sum_{f \in \mathrm{flavors}} \bar{\psi}_f \left(-i\partial\!\!\!/ + m_f + g V\!\!\!/ + i\mu_f\gamma_4 \right) \psi_f + \sum_{f \in \mathrm{flavors}} \bar{\psi}_f g X\!\!\!/ \psi_f \right], \tag{6.2.1}$$

---

[14]Note $X_\mu(x) \times X_\mu(x) = 0$.

and the gauge part, derived from the Yang-Mills Lagrangian, is given in Eq. (6.1.30):

$$
\begin{aligned}
\mathcal{S}_{\text{gauge}} = \int \mathrm{d}^4x \Bigg[&\frac{1}{4}\Big[(F_{3,\mu\nu} + H_{3,\mu\nu})^2 + (F_{8,\mu\nu} + H_{8,\mu\nu})^2\Big] + \frac{1}{2} X_{a,\mu} \mathcal{Q}^{ab}_{\mu\nu} X_{b,\nu} + \\
&-\frac{1}{2}\left(D_\mu[V]X_\nu - D_\nu[V]X_\mu\right)\cdot g(X_\mu \times X_\nu) + \frac{1}{4}g^2(X_\mu \times X_\nu)^2 \Bigg] .
\end{aligned}
\tag{6.2.2}
$$

When dealing with $\hat{n}_i(x), c_{i,\mu}(x)$ and $X_\mu(x)$ as *independent* variables, we have to make sure that $\hat{n}_i(x)\cdot\hat{n}_i(x) = 1$ and $\hat{n}_i(x)\cdot X_\mu(x) = 0$. These two constraints are invariant under the gauge transformations (6.1.37) by construction. The most general method to include the gauge-fixing condition is provided by the Becchi-Rouet-Stora-Tyutin (BRST) quantization, a generalization of the Faddeev-Popov quantization. Applying the BRST quantization to the CFNS theory by implementing the gauge-fixing condition (6.1.39), the remaining SU(3) gauge symmetry is fully retained.

The BRST quantization proceeds in the same way as the Fadeev-Popov method,[15] i.e., one writes down the generating functional, $\mathcal{Z}_{\text{CFNS}}$, including the constraints as delta functions imposed on the fields. In our case we obtain specifically[16]

$$
\mathcal{Z}_{\text{CFNS}} = \int \mathcal{D}\hat{n}\,\mathcal{D}c\,\mathcal{D}X \; J \, \exp\left(-\mathcal{S}^{\text{QCD}}_{\text{E}}[\hat{n}, c, X]\right) ,
\tag{6.2.3}
$$

where the components of the fields, over which the integration has to be extended, are collected in $\hat{n} \equiv \{\hat{n}_i\}, c \equiv \{c_{i,\mu}\}, X \equiv \{X_\mu\}$. The functional determinant J appears when changing the variables of the extended gauge theory from $(\hat{n}, A)$ to $(\hat{n}, c, X)$. It turns out to be $J = 1$ (see Ref. [149]), and will hence be neglected. In the standard BRST quantization scheme, which can be found in the literature (e.g., Ref. [170]), the gauge-fixing condition is implemented by introducing, first, ghost and antighost fields G and $\bar{G}$, respectively, plus the so-called Nakanishi-Lautrup auxiliary fields N and $\bar{N}$. The additional contribution to the action, $\mathcal{S}_{\text{FP+GF}}$, that describes both the (generalized) Faddeev-Popov ghosts and antighosts, and the gauge-fixing condition, is determined by

$$
\mathcal{S}_{\text{FP+GF}} = -\mathrm{i} \int \mathrm{d}^4x\, \delta_{\text{BRST}}\big(\bar{G}\cdot \chi[\hat{n}, C_\mu, X_\mu]\big) ,
\tag{6.2.4}
$$

where $\delta_{\text{BRST}}(Y)$ denotes the BRST transformation of the field $Y \in \{\hat{n}, c, X, G, \bar{G}, N, \bar{N}\}$ corresponding to the gauge transformation (6.1.37). This transformation is characterized by being nilpotent, i.e., $\delta^2_{\text{BRST}} = 0$. Since its construction is rather involved, we refer the interested reader to the literature, cf. Ref. [171], and content ourself here with the result

$$
\mathcal{S}_{\text{FP+GF}} = \int \mathrm{d}^4x \Big[N\cdot D_\mu[V]X_\mu + \mathrm{i}\bar{G}\cdot D_\mu[V]D_\mu[A]G + g^2\left(\mathrm{i}\bar{G}\cdot(\hat{n}\times X_\mu)\right)\left((\hat{n}\times X_\mu)\cdot G\right) \Big] .
\tag{6.2.5}
$$

The full generating function, including the correct gauge-fixing condition in order to eliminate

---

[15]The advantage of the BRST over the Faddeev-Popov quantization is that the first allows an arbitrary gauge-fixing condition while the latter is restricted to gauge-fixing conditions that can be expressed in terms of ghost-antighost terms.

[16]Here, we suppress the source term, because it is not needed.

the redundant $\mathrm{SU}(3)/(\mathrm{U}(1) \times \mathrm{U}(1))$ local symmetry, is then given by

$$\tilde{\mathcal{Z}}_{\mathrm{CFNS}} = \int \mathscr{D}\hat{n} \,\mathscr{D}c \,\mathscr{D}X \,\mathscr{D}G \,\mathscr{D}\bar{G} \,\mathscr{D}N \,\mathscr{D}\bar{N} \, \exp\left[-\left(\mathcal{S}_{\mathrm{E}}^{\mathrm{QCD}}[\hat{n}, c, X] + \mathcal{S}_{\mathrm{FP+GF}}[G, \bar{G}, N, \bar{N}, \hat{n}, c, X]\right)\right]. \tag{6.2.6}$$

Carrying out the functional integrals over the fields G, N and X_μ, keeping only terms up to second order ("one-loop level") in $\mathcal{S}_{\mathrm{E}}^{\mathrm{QCD}}$ and $\mathcal{S}_{\mathrm{FP+GF}}$ and completing the square with respect to X_μ in $\mathcal{S}_{\mathrm{E}}^{\mathrm{QCD}}$, the effective CFNS action results as follows:

$$\begin{aligned}\mathcal{S}_{\mathrm{CFNS}} = &\int \mathrm{d}^4x \left[\frac{1}{4}\hat{G}_{\mu\nu}^2 + \sum_{f \in \mathrm{flavors}} \bar{\psi}_f \left(-\mathrm{i}\slashed{\partial} + m_f + g\slashed{V} + \mathrm{i}\mu_f\gamma_4\right)\psi_f\right] + \\ &+ \int \mathrm{d}^4x \,\mathrm{d}^4y \,\frac{1}{2} \left(\bar{\psi}(x)\gamma_\mu t_a \psi(x)\right) g^2 \left(\mathcal{Q}^{-1}\right)_{\mu\nu}^{ab}(x, y)\left(\bar{\psi}(y)\gamma_\nu t_b \psi(y)\right) + \\ &+ \frac{1}{2}\ln\det \mathcal{Q}_{\mu\nu}^{ab} - \ln\det \mathcal{R}^{ab},\end{aligned} \tag{6.2.7}$$

with $\hat{G}_{\mu\nu}^c$ defined in Eq. (6.1.24), and (cf. Eq. (6.1.31))

$$\mathcal{Q}_{\mu\nu}^{ab} = \mathcal{R}^{ab}\delta_{\mu\nu} + 2g f_{abc}\hat{G}_{\mu\nu}^c, \qquad \mathcal{R}^{ab} := -\left(D_\rho[V]D_\rho[V]\right)^{ab}. \tag{6.2.8}$$

The term in the second line comes from the completion of the square, the first term in the third line comes from the integration over X_μ in Eq. (6.1.30), and the last term in the third line is the Faddeev-Popov determinant[17]. Here, $(\mathcal{Q}^{-1})_{\mu\nu}^{ab}$ is the X-field correlator and a nonlocal four-fermion interaction is generated. This is exactly the expression that appears in the nonlocal PNJL model. The range of the nonlocality is determined by the correlation length ξ, characteristic of the color exchange through gluon fields, and it is given by the inverse M_X (defined in the previous section), thus $\xi \sim M_X^{-1} \approx 0.15\,\mathrm{fm}$.

6.2.1 Weiss Potential and Renormalization Group Equation

After the determination of the effective action, $\mathcal{S}_{\mathrm{CFNS}}$, in Eq. (6.2.7), we can now proceed with its further evaluation. First, consider the gauge part,

$$\mathcal{S}_{\mathrm{CFNS}}^{\mathrm{gauge}} = \int \mathrm{d}^4x \,\frac{1}{4}\hat{G}_{\mu\nu}^2 + \frac{1}{2}\ln\det \mathcal{Q}_{\mu\nu}^{ab}. \tag{6.2.9}$$

We adopt the Polyakov gauge for the four-component of the field $V_\mu(x)$:

$$V_{a,4}(x) = c_{3,4}(\vec{x})\delta_{a3} + c_{8,4}(\vec{x})\delta_{a8} \quad \Longrightarrow \quad \partial_4 V_{3,4}(x) = 0, \quad \partial_4 V_{8,4}(x) = 0. \tag{6.2.10}$$

Notice, however, that the Polyakov gauge, Eq. (6.2.10), is not complete. An additional gauge condition must be chosen in order to eliminate all nonphysical degrees of freedom from the theory. Here we use the Abelian gauge, fixing $\hat{n}_3(x) \equiv \hat{e}_3 := (0,0,1,0,0,0,0,0)^\top$ and $\hat{n}_8(x) \equiv \hat{e}_8 := (0,0,0,0,0,0,0,1)^\top$ in adjoint representation. The advantage of this gauge fixing is that

---

[17]This term follows from Eq. (6.2.5) after writing $\bar{G}\cdot D_\mu[V]D_\mu[A]G = \bar{G}\cdot D_\mu[V]D_\mu[V]G + \bar{G}\cdot D_\mu[V]gX_\mu\times G$ and neglecting the last expression because being cubic in the ghost and $X_\mu(x)$ field.

the restricted connection, $V_\mu(x)$, is proportional to the Cartan-algebra elements t_3, t_8. This, in turn, reduces the integration in the generating functional (6.2.6) over the fields c. Next, decompose the gluon part,[18]

$$\frac{1}{4}\hat{G}_{\mu\nu}^2 = \frac{1}{2}\hat{G}_{4j}^2 + \frac{1}{4}\hat{G}_{jk}^2 ,$$

into "electric" and "magnetic" pieces. The electric part is

$$\frac{1}{2}\hat{G}_{4j}^2 = \frac{1}{2}\left(\partial_j V_{a,4}\right)^2 - \frac{1}{2}V_{a,j}D_4^{ac}[V]D_4^{cb}[V]V_{b,j} , \tag{6.2.11}$$

with $D_4^{ac}[V]D_4^{cb}[V]$ given in Eq. (6.1.32),

$$\begin{aligned}
-D_4^{ac}[V]D_4^{cb}[V] = &-\delta^{ab}\partial_4^2 - 2gf_{3ab}V_{3,4}\partial_4 - 2gf_{8ab}V_{8,4}\partial_4 + \\
&+ g^2 V_{3,4}V_{3,4}\text{diag}\left(1,1,0,\frac{1}{4},\frac{1}{4},\frac{1}{4},\frac{1}{4},0\right)_{ab} + \\
&+ g^2 V_{3,4}V_{8,4}\text{diag}\left(0,0,0,\frac{\sqrt{3}}{2},\frac{\sqrt{3}}{2},-\frac{\sqrt{3}}{2},-\frac{\sqrt{3}}{2},0\right)_{ab} + \\
&+ g^2 V_{8,4}V_{8,4}\text{diag}\left(0,0,0,\frac{3}{4},\frac{3}{4},\frac{3}{4},\frac{3}{4},0\right)_{ab} .
\end{aligned} \tag{6.2.12}$$

The magnetic part is

$$\frac{1}{4}\hat{G}_{jk}^2 = \frac{1}{2}V_{a,j}\left(-\delta_{jk}\partial_\ell\partial_\ell + \partial_j\partial_k\right)V_{a,k} + \mathcal{O}(V^3) . \tag{6.2.13}$$

Next, we expand the theory around the nontrivial background for V_4 and assume vanishing expectation values for the spatial components V_j:

$$\begin{aligned}
V_{a,4}(x) &= 2\varphi_1/(g\beta m_3)\delta^{a3} + v_{3,4}(\vec{x})\delta^{a3} + 2\varphi_2/(g\beta m_8)\delta^{a8} + v_{8,4}(\vec{x})\delta^{a8} \\
V_{a,j}(x) &= 0 + v_{a,j}(x) = c_{3,j}(x)\delta^{a3} + c_{8,j}(x)\delta^{a8} ,
\end{aligned} \tag{6.2.14}$$

where the last line is valid for the Abelian gauge assumed here. Recall that $\vec{m} := (m_3, m_8)$ denotes the maximal weight vector of a representation. The prefactors in the expectation value of $V_{a,4}$ are chosen such that the Polyakov loop Φ, calculated using Eq. (6.1.36), is given in leading order in V_4 by[19]

$$\Phi = \frac{1}{3}\text{tr}_c\left[\exp\left(\text{i}\varphi_1 t_3 + \text{i}\varphi_2 t_8\right)\right] . \tag{6.2.15}$$

From the explicit form of the magnetic part, Eq. (6.2.13), of the field strength tensor it is convenient to split the spatial fluctuations, $v_{a,j}$, into the transverse and longitudinal components, $v_{a,j}^t, v_{a,j}^l$, respectively,

$$v_{a,j}^t(x) := \Pi_{jk}^t v_{a,k}(x) := \left(\delta_{jk} - \frac{\partial_j\partial_k}{\partial_\ell\partial_\ell}\right)v_{a,k}(x)$$

$$v_{a,j}^l(x) := \Pi_{jk}^l v_{a,k}(x) := (1 - \Pi_{jk}^t)v_{a,k}(x) ,$$

---

[18]Latin characters denote the spatial components, only, i.e., $j \in \{1, 2, 3\}$.
[19]Note the change in notation compared to Eq. (5.2.17): for convenience, we have set $\varphi_i = 2\phi_i$ for $i \in \{3, 8\}$.

with the transverse and longitudinal projection operators, Π^t, Π^l, which satisfy $(\Pi^t)^2 = (\Pi^l)^2 = 1$ and $\Pi^t \Pi^l = 0$. Then the gauge part of the CFNS action (6.2.9) up to quadratic order in the fluctuations, reads:

$$S_{\text{CFNS}}^{\text{gauge}} = \frac{1}{2} \ln \det \mathcal{Q}_{\mu\nu}^{ab} + \frac{\beta}{2} \int d^3\vec{x} \left[v_{3,4}(\vec{x})(-\partial_\ell \partial_\ell) v_{3,4}(\vec{x}) + v_{8,4}(\vec{x})(-\partial_\ell \partial_\ell) v_{8,4}(\vec{x}) \right] +$$
$$+ \frac{1}{2} \int d^4x \, v_{a,j}^t(x) \left\{ -\delta^{ab} \partial_\ell \partial_\ell - D_4^{ac}[V] D_4^{cb}[V] \right\} v_{b,j}^t(x) + \qquad (6.2.16)$$
$$+ \frac{1}{2} \int d^4x \, v_{a,j}^l(x) \left\{ -D_4^{ac}[V] D_4^{cb}[V] \right\} v_{b,j}^l(x) + \mathcal{O}(v^3) \,.$$

Now, the functional integration over the fields v_4, v_j^t, v_j^l can be performed[20] with $-D_4^{ac}[V]D_4^{cb}[V]$ evaluated at the mean-field values $V_{3,4}(x) \to 2\varphi_1/(g\beta m_3), V_{8,4}(x) \to 2\varphi_2/(g\beta m_8)$, leading to

$$S_{\text{CFNS}}^{\text{gauge}} = \frac{1}{2} \ln \det \mathcal{Q}_{\mu\nu}^{ab} + 2 \cdot \frac{1}{2} \text{Tr} \ln(-\partial_\ell \partial_\ell) + 2 \cdot \frac{1}{2} \text{Tr} \ln \left\{ -\delta^{ab} \partial_\ell \partial_\ell - D_4^{ac}[V] D_4^{cb}[V] \right\} +$$
$$+ \frac{1}{2} \text{Tr} \ln \left\{ -D_4^{ac}[V] D_4^{cb}[V] \right\} \,.$$

Observing $\mathcal{Q}_{\mu\nu}^{ab} = \delta_{\mu\nu} \mathcal{R}^{ab} = -\delta_{\mu\nu} D_4^{ac}[V] D_4^{cb}[V]$ in Polyakov gauge, the first term on the right-hand side (from the integration over the fields $X_\mu(x)$) and the first term in the second line (from the longitudinal gluons) is exactly canceled by the Faddeev-Popov ghost term, $-\ln \det \mathcal{R}^{ab}$, that contributes to the total action (6.2.7). Therefore, the gauge part of the action reduces to

$$S_{\text{CFNS}}^{\text{gauge}} \to \tilde{S}_{\text{CFNS}}^{\text{gauge}} = \text{Tr} \ln(-\partial_\ell \partial_\ell) + \text{Tr} \ln \left\{ -\delta^{ab} \partial_\ell \partial_\ell - D_4^{ac}[V] D_4^{cb}[V] \right\} \,. \qquad (6.2.17)$$

Finally, using $\Omega_{\text{CFNS}}^{\text{gauge}} = \frac{T}{V} \tilde{S}_{\text{CFNS}}^{\text{gauge}}$ and exploiting the Matsubara formalism, the traces can be evaluated, leading to the following (φ_1, φ_2)-dependent contribution to the free energy $\Omega_{\text{CFNS}}^{\text{gauge}}$:

$$\Omega_{\text{CFNS}}^{\text{gauge}} \equiv \mathcal{V}_W(\varphi_1, \varphi_2) = f(\varphi_1) + f\left(\frac{1}{2}\varphi_1 + \frac{\sqrt{3}}{2}\varphi_2 \right) + f\left(-\frac{1}{2}\varphi_1 + \frac{\sqrt{3}}{2}\varphi_2 \right) \,, \qquad (6.2.18)$$

with

$$f(\varphi) = \frac{1}{2} T \int \frac{d^3\vec{p}}{(2\pi)^3} 4 \ln \left[1 - 2e^{-\beta p} \cos\varphi + e^{-2\beta p} \right] \qquad (6.2.19)$$
$$\simeq T^4 \left[-\frac{1}{6}(\varphi - \pi)^2 + \frac{1}{12\pi^2}(\varphi - \pi)^4 + \frac{\pi^2}{12} \right] \quad (\text{mod } 2\pi) \,.$$

$\mathcal{V}_W$ is the so-called Weiss potential (cf. [172]). It exhibits $Z(3)$ symmetry. It turns out that the minimum of the Weiss potential is realized at the point where the $Z(3)$ symmetry is spontaneously broken (e. g., $(\varphi_1, \varphi_2) = (0,0), (\varphi_1, \varphi_2) = (0, 4\pi/\sqrt{3}), (\varphi_1, \varphi_2) = (2\pi, 2\pi/\sqrt{3})$). Thus the

---

[20]Note, that the integration over $\hat{n}, c$ has been changed into an integration over v_4, v_j^t, v_j^l. Not all of the latter fields are, hence, independent and one would require an additional Faddeev-Popov determinant term in order to get rid of the unphysical degrees of freedom. We will see, however, that the ghost term $-\ln \det \mathcal{R}^{ab}$ in Eq. (6.2.7) already removes the (unphysical) longitudinal contributions. The reason for this is that the reduction scheme used for breaking down the enlarged CFNS symmetry (6.1.23) to the local SU(3) symmetry was already an Abelian gauge-fixing condition similar to that imposed here on $\hat{n}(x)$.

Weiss potential invariably describes the deconfined (high-temperature) phase. The treatment of the action up to second order in the fluctuations of the gauge field $V_\mu(x)$ only reproduces the high-temperature case where a perturbative expansion applies. The Weiss potential is generated by the spatial fluctuations of the transverse components of the field $V_\mu(x)$. In turn, one concludes that (in Polyakov gauge) fluctuations of the four-component, $V_4(x)$, carry the confining properties of the Polyakov-loop variables.

The next step is to include fluctuations of $V_4(x)$ in order to obtain a Polyakov-loop potential that describes the confinement-deconfinement transition in pure Yang-Mills theory. As shown in Ref. [173], the effective Polyakov-loop potential can best be extracted by calculating the effective action Γ within Wetterich's functional renormalization-group approach. To this end, one introduces an infrared cutoff for the transversal spatial gauge fields and in the temporal gauge fields by modifying the action $\mathcal{S}_{\text{CFNS}}^{\text{gauge}}$ according to $\mathcal{S}_{\text{CFNS}}^{\text{gauge}} \to \mathcal{S}_{\text{CFNS}}^{\text{gauge}}[v_4, v^t] + \Delta \mathcal{S}_k[v_4] + \Delta \mathcal{S}_k^t[v^t]$, with an infrared scale k and cutoff terms

$$\begin{aligned} \Delta \mathcal{S}_k[v_4] &= \frac{1}{2}\beta \int \mathrm{d}^3\vec{x}\, v_4 R_{4,k} v_4 \\ \Delta \mathcal{S}_k^t[v^t] &= \int_0^\beta \mathrm{d}\tau \int \mathrm{d}^3\vec{x}\, v_{a,i}^t R_k^t v_{a,i}^t \,. \end{aligned} \qquad (6.2.20)$$

The regulators R_k are chosen to be momentum dependent and required to provide masses at low momenta and to vanish at large momenta. For $k \to 0$ they vanish identically, and one recovers the full nonperturbative result for the action Γ.[21]

The flow of the cutoff-dependent effective Γ_k is then given by Wetterich's equation, cf. Refs. [174–176], for Yang-Mills theory in Polyakov gauge,

$$\partial_t \Gamma_k[v_4, v^t] = \frac{1}{2}\mathrm{Tr}\left\{ \left(\Gamma_k^{(2)}[v_4, v^t] + R_k\right)^{-1} \partial_t R_k \right\}, \qquad (6.2.21)$$

where $t := \ln(k/\Lambda)$, and Λ is some reference scale. $\left(\Gamma_k^{(2)}\right)_{\mu\nu} := \frac{\partial^2 \Gamma_k}{\partial v_\mu \partial v_\nu}$ denotes the second derivatives (propagator) of the effective action with respect to the fields $v_4, v_j := v_j^t$. In Polyakov gauge, it turns out (cf. Ref. [173]), that the simple approximation calculated above in Eq. (6.2.17), is sufficient to describe the confinement-deconfinement phase transition. Hence, we have

$$\begin{aligned} \Gamma_k[v_4, v^t] = \beta \int \mathrm{d}^3\vec{x}\left[\frac{1}{2}v_{3,4}(\vec{x})(-Z_{4,k}\partial_\ell\partial_\ell)v_{3,4}(\vec{x}) + \frac{1}{2}v_{8,4}(\vec{x})(-Z_{4,k}\partial_\ell\partial_\ell)v_{8,4}(\vec{x}) + V_k[v_4]\right] + \\ + \frac{1}{2}\int \mathrm{d}^4x\, Z_{j,k}v_{a,j}^t(x)\left\{-\delta^{ab}\partial_\ell\partial_\ell - D_4^{ac}[V]D_4^{cb}[V]\right\}v_{b,j}^t(x)\,, \end{aligned}$$

$$(6.2.22)$$

with k-dependent wave-function renormalizations $Z_{4,k}, Z_{j,k}$. The effective Polyakov potential,

---

[21]We do not specify the explicit form of the regulators R_k here. We mention, however, that $R_k(q)$ should satisfy three requirements: First, it should play the role of an infrared regularization such that $\lim_{q^2/k^2 \to 0} R_k(q) > 0$, i.e., the regulator term can, indeed, be viewed as a momentum-dependent mass term. Second, one demands $\lim_{k^2/q^2 \to 0} R_k(q) = 0$, i.e., one automatically recovers the standard generating functional, as well as the full effective action in this limit: $\lim_{k \to 0} \mathcal{Z}_k = \mathcal{Z}$ and $\lim_{k \to 0} \Gamma_k = \Gamma$. Third, the functional integral is dominated by the stationary point of the action in the limit $k \to \infty$: $\lim_{k^2 \to \Lambda^2 \to \infty} R_k(q) \to \infty$. This justifies the use of the saddle-point approximation which filters out the classical field configuration and the bare action, $\Gamma_{k \to \Lambda} \to \mathcal{S} + \text{const.}$, where Λ is some reference scale.

$V_{\text{eff}}[v_4]$, is given by $V_{\text{eff}}[v_4] := \lim_{k\to 0} V_k[v_4]$, and $V_k[v_4]$ has to be determined via the Wetterich flow equation (6.2.21). Here we do not go into all details of the calculation, and refer the reader to Ref. [173]. We comment, however, on some of the keystones of the calculation: using the truncation scheme of Ref. [173], leading to Eq. (6.2.22), and neglecting back-reactions of $V_k[v_4]$ on the transversal gauge fields, the flow equations (6.2.21) for the temporal, v_4, and spatial, v^t, components decouple. This allows one to split the effective potential into the corresponding pieces, $V_k[v_4] := V_k^t[v_4] + \Delta V_k[v_4]$. From the calculations above it is clear that $\lim_{k\to 0} V_k^t[v_4] = \mathcal{V}_W(v_{3,4}, v_{8,4})$, i.e., it reproduces the Weiss potential.

We are finally left with the determination of ΔV_k, the part of the effective potential induced by V_4 fluctuations. Since the contributions from transversal fields have been treated independently of those of the temporal fields, it suffices to consider the following action:

$$\Gamma_k[v_4] = -\beta \int d^3\vec{x} \left[\frac{1}{2} v_{3,4}(\vec{x}) Z_{4,k} \partial_\ell \partial_\ell v_{3,4}(\vec{x}) + \frac{1}{2} v_{8,4}(\vec{x}) Z_{4,k} \partial_\ell \partial_\ell v_{8,4}(\vec{x}) - \Delta V_k[v_4] - V_k^t[v_4] \right].$$
(6.2.23)

This action allows one to derive the flow equation for the potential generated by the fluctuations of V_4,

$$\beta \partial_t \Delta V_k = \frac{1}{2} \int \frac{d^3\vec{p}}{(2\pi)^3} \frac{\partial_t R_{4,k}}{Z_4 \vec{p}^2 + \partial_4^2 (\Delta V_k + V_k^t) + R_{4,k}}.$$
(6.2.24)

The resulting effective action, $V_{\text{eff}}[v_4] = \lim_{k\to 0} (V_k^t[v_4] + \Delta V_k[v_4])$, is shown in Fig. 6.1 for the SU(2) case. This potential leads to a critical temperature $T_c^{\text{SU}(2)} = 305^{+40}_{-55}$ MeV [173] (compare also Ref. [177]).

Summarizing the Yang-Mills part, we find an effective Polyakov potential that describes the confinement-deconfinement transition when both, the transverse- and temporal-gluon fluctuations are included. Transverse gluons, v^t, govern the perturbative behavior in the deconfined high-temperature phase while fluctuations of the temporal gluons, v_4, are responsible for the confinement.

6.2.2 Derivation of the Nonlocal PNJL Model

After the discussion of the Yang-Mills part of the reformulated (CFNS) quantum chromodynamics, we are now ready to derive the nonlocal PNJL model. Let us first address the Polyakov-loop effective potential. QCD in the one-loop approximation cannot account for confinement: the Weiss potential (6.2.18) leads to deconfinement at all temperatures. Only the full nonperturbative treatment, using the renormalization group equations generates confinement. The one-loop treatment shows, however, that a potential describing deconfinement must include transverse gluons as degrees of freedom. It can therefore be stated that the Polyakov-loop effective potential, $\mathcal{U}$ of Eq. (5.2.20), used in our calculations and leading to the correct Stefan-Boltzmann limit, already incorporates transverse gluons parametrized in terms of the temporal field v_4.

Having clarified the origin of the $\mathcal{U}$ in the nonlocal PNJL model, we can now relate the four-fermion interaction in the CFNS action (6.2.7) to that of the nonlocal PNJL model (Eq. (3.2.4)). Under the assumption of $(\mathcal{Q}^{-1})_{\mu\nu}^{ab}$ being diagonal in color space and Euclidean time-space, it is easy to recover the following relation between the momentum distribution function, $\mathcal{C}(z)$, of the

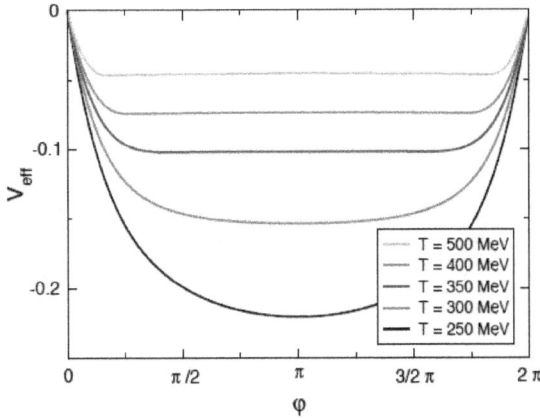

FIGURE 6.1: Full effective SU(2) potential V_{eff} as a function of the (normalized) temporal component of the gluon field, $\varphi := \beta g v_4$. The differently colored lines correspond to temperatures ranging from 500 MeV in the deconfined region to 250 MeV in the confined phase. The expectation value $\langle\varphi\rangle$ (that determines the Polyakov loop $L[\langle v_4\rangle] = \cos\left(\frac{\langle\varphi\rangle}{2}\right)$) in the center-broken deconfined phase is given by the transition point between the decreasing part of the potential for small φ and the flat region in the middle of the plot. In the center-symmetric confined phase it is just given by the minimum at $\varphi = \pi$ (from Ref. [173]).

nonlocal PNJL model, and the X-field correlator $(\mathcal{Q}^{-1})^{ab}_{\mu\nu}$ (in Euclidean time-space):

$$g^2 \left(\mathcal{Q}^{-1}\right)^{ab}_{\mu\nu}(x,y) = G\delta_{\mu\nu}\delta_{ab}\mathcal{C}(x-y)\,,$$

($g^2 = 4\pi\alpha_{\mathrm{s}}$ denotes the strong-interaction coupling strength) from which we obtain

$$\begin{aligned} G\,\mathcal{C}(x-y) &= g^2 \left(\mathcal{Q}^{-1}\right)^{ab}_{\mu\nu}(x,y)\frac{\delta_{\mu\nu}}{4}\frac{\delta_{ab}}{N_c^2-1} \\ &= g^2 \frac{\mathrm{Tr}\left(\mathcal{R}^{-1}\right)}{N_c^2-1}\,. \end{aligned} \tag{6.2.25}$$

Changing to Euclidean momentum space, the explicit calculation of the inverse correlator, $(\mathcal{R}^{-1})^{ab}$, in Polyakov gauge (we have $\hat{G}^c_{\mu\nu} = 0$, and thus $\mathcal{Q}^{ab}_{\mu\nu}$ is determined by Eq. (6.1.32) with Eq. (6.2.12)) leads to

$$\begin{aligned} \mathrm{Tr}\left(\mathcal{R}^{-1}\right)_{\omega,\vec{p}} = {}& 2\phi_{\omega,\vec{p}}(0,0) + \phi_{\omega,\vec{p}}(\varphi_1,0) + \phi_{\omega,\vec{p}}(-\varphi_1,0) + \\ & + \phi_{\omega,\vec{p}}\left(\frac{1}{2}\varphi_1, \frac{\sqrt{3}}{2}\varphi_2\right) + \phi_{\omega,\vec{p}}\left(-\frac{1}{2}\varphi_1, \frac{\sqrt{3}}{2}\varphi_2\right) + \\ & + \phi_{\omega,\vec{p}}\left(\frac{1}{2}\varphi_1, -\frac{\sqrt{3}}{2}\varphi_2\right) + \phi_{\omega,\vec{p}}\left(-\frac{1}{2}\varphi_1, -\frac{\sqrt{3}}{2}\varphi_2\right)\,, \end{aligned} \tag{6.2.26}$$

where

$$\phi_{\omega,\vec{p}}(\varphi_1,\varphi_2) := \frac{1}{\vec{p}^2 + (\omega - (\varphi_1 + \varphi_2)/\beta)^2} \, . \tag{6.2.27}$$

With this, we obtain exactly the momentum distribution function, Eq. (3.2.39) (compare, also, Eqs. (5.3.6) and (5.3.14)), that we used in our finite-temperature calculations, once $g^2 = 4\pi\alpha_s$ is used.

In summary, the nonlocal PNJL model used in the present work has indeed its basis in a consistent QCD treatment. The form of the nonlocality distribution function $\mathcal{C}(p)$ derives from this formulation. It has been demonstrated that the Polyakov loop depends only on the fields $V_\mu(x)$, i.e., the gauge fields that lie in the Cartan algebra of the SU(3) Lie group. An expansion around nonvanishing expectation values of their temporal components, $V_4(x)$, and assuming vanishing spatial components $\langle V_j(x) \rangle = 0$, $j \in \{1,2,3\}$ allows one to recover, first, a deconfining (Weiss) potential governed by fluctuations of the transverse fields v^t, and then, using the renormalization group equations, a confining potential directed by the fluctuations of the temporal components v_4. We conclude for the gauge part that all this information is incorporated in the Polyakov-loop effective potential $\mathcal{U}$ used in the Polyakov-loop-extended NJL model. Concerning the matter part, we have obtained the nonlocal four-fermion interaction with a distribution function of the form used in the PNJL model and previously motivated by the operator product expansion. With this we conclude our extended derivation of the nonlocal PNJL model from QCD first principles.

7 Discussion and Conclusion

In this work we have derived and presented a nonlocal Polyakov-loop-extended Nambu–Jona-Lasinio model, suitable for a description of strongly interacting matter at finite temperature and finite density. In our presentation, we have followed the "historical", bottom-up development of the model. We have started with the local NJL models, pointed out their limitations (imposed, in particular, by the characteristic momentum cutoff), and have then extended the model by establishing contact with Dyson-Schwinger calculations. The resulting model includes a nonlocal four-fermion coupling plus a six-fermion Kobayashi-Maskawa-'t Hooft interaction. It has turned out, that this nonlocal NJL model can be coupled to the Polyakov loop, the order parameter for the confinement-deconfinement transition in the pure gauge theory. Given this framework we have derived the phase diagrams as they result from both the two- and three-flavor nonlocal PNJL models.

7.1 Discussion

At this stage it is worth pointing out that we have actually gained more than yet another model for a tentative description of the phase diagram of QCD. Based on and extending recent work by K.-I. Kondo, we have clarified in Chapt. 6 how the $N_f = 3$ nonlocal PNJL model can be derived from first QCD principles. The reformulation of quantum chromodynamics in terms of the Cho-Faddeev-Niemi-Shabanov decomposition of the gauge fields allows one to consistently derive the PNJL model and to understand the impact of the assumptions and simplifications made in constructing the PNJL model. The role played by the gauge sector has been investigated in detail. In lattice calculations it has been shown that the nondiagonal SU(3) gauge fields become massive valence gluons. It follows that only the diagonal fields should play a role when describing confinement. Indeed, the Polyakov loop can be parametrized in terms of the SU(3)-Cartan-algebra elements A^3 and A^8 only. We have clarified the impact of transverse gluons: at first sight one might assume that the Polyakov-loop-effective potential $\mathcal{U}$, used in the present work, includes only the contributions from longitudinal gluons since it exclusively depends on A_4^3 and A_4^8. We have shown, however, that an effective potential describing deconfinement implicitly incorporates contributions from transverse gluons. Finally, treating the fermion sector of the reformulated QCD allows one to recover exactly the nonlocal PNJL model used in this work.

In essence we now have a solid basis for and a profound understanding of (non)local PNJL models. The derivation of the nonlocal PNJL model starting from full QCD allows one to systematically include higher-order corrections.

Turning the discussion to thermodynamics, we have first considered the two-flavor scenario.

The strong entanglement of the chiral and the confinement-deconfinement transitions is manifest. It is demonstrated to persist for the three-flavor case as well. Several thermodynamical quantities have been calculated, such as the pressure, the energy density and the sound velocity, at the mean-field level and including second-order corrections from pions.

The major purpose of the two-flavor studies was a direct comparison to two-flavor calculations performed in the local model [38]. The results up to the critical temperature, $T \sim T_c$, are very similar, but above T_c, the results differ reflecting the impact of the momentum cutoff in the local PNJL model. In a second step, the same thermodynamical quantities have been calculated for the three-flavor case. Furthermore, we have determined the QCD phase diagram as it results from the nonlocal PNJL model. The three-flavor model is more involved compared to the two-flavor case, because of the axial anomaly. In particular, we have calculated the movement of the critical point as a function of the strength of the axial anomaly. In contrast to local-PNJL investigations we have found that the critical point moves considerably on the chemical-potential axis, while it is more stable on the temperature axis. This behavior can be understood from the Columbia plot (cf. Chapt. 4) taking into account the particular values of parameters used in the nonlocal PNJL model: in order to reproduce the physical meson spectrum, we have used current quark masses $m_u = m_d = 3\,\text{MeV}$ and $m_s = 70\,\text{MeV}$ (corresponding to a renormalization scale of about $2\,\text{GeV}$). On the other hand, the axial-anomaly strength in the nonlocal model had to be chosen about twice as large compared to the local model. From local-PNJL model calculations [178] we know that the first-order chiral-transition region in the Columbia plot is increasing with a rising anomaly strength. It is clear that the axial anomaly has a major impact on the structure of the QCD phase diagram.

Furthermore, when considering the phase diagram of QCD, we have noticed that the first-order transition line ends in nuclear territory when $T \to 0$. This clearly indicates that important degrees of freedom are missing when approaching nuclear-matter densities and low temperatures. Regardless of whether the first-order line ends above or below $\mu_B = M_N = 940\,\text{MeV}$, it must always be kept in mind that the known physics of nuclear matter influences the phase structure qualitatively.

We have conceded some space to the discussion of finite-temperature lattice results. In the last four years, major progress has been made in calculations at finite temperature but zero density. It appears to be well-established that the transition from the chirally broken to the chirally restored phase is a crossover. The actual challenge is, however, to determine the transition temperature T_c itself. Until a year ago two numbers, $T_c^{\text{BMW}} \approx 150\,\text{MeV}$ of the Budapest-Marseille-Wuppertal collaboration, and $T_c^{\text{"hotQCD"}} \approx 190\,\text{MeV}$ of the "hotQCD" collaboration, existed that could not be reconciled with one another. During this year 2010 the "hotQCD" collaboration has investigated the impact of discretization errors and of unphysically large meson masses on their results. As of today it seems most likely that both collaborations converge towards a transition-temperature range of $T_c \approx 155\text{--}170\,\text{MeV}$.

Despite open questions regarding the lattice results, we show in Fig. 7.1 our nonlocal PNJL-model results for the chiral condensate including pionic corrections (compare Fig. 5.12). Thermal pions, owing to their small masses, influence the behavior of the chiral condensate in the tran-

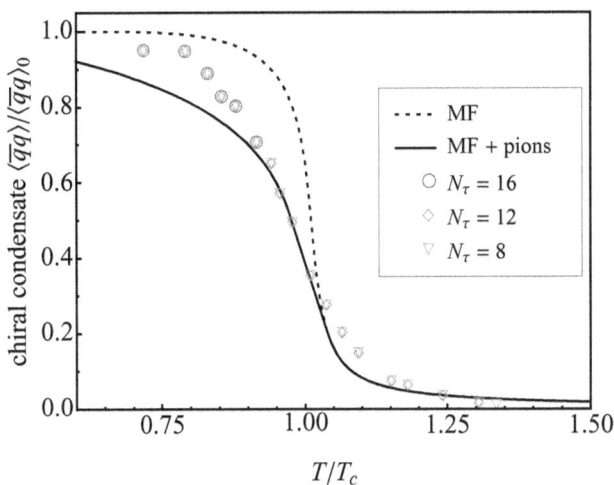

FIGURE 7.1: Chiral condensate calculated in mean-field approximation (dashed line) and including thermal pions (solid line) compared to lattice data of the Budapest-Wuppertal-Marseille collaboration using physical meson masses [99].

sition region (note, that $T_c^{\mathrm{BMW}} \approx 150\,\mathrm{MeV} \sim m_\pi$). It will be important in future investigations to properly include the effects induced by pions and other light mesons.

7.2 Conclusion

The aim of this work has been to construct a model beyond the local NJL approach with its simplistic momentum-space cutoff. Basic features related to the chiral and confinement-deconfinement transitions are well reproduced. Further steps must include diquark degrees of freedom. From a schematic Ginzburg-Landau treatment [179, 180] it turns out that diquarks in color superconductors play an essential role in the low-temperature and high-density region. When including a diquark-diquark-chiral condensate interaction, it is even possible to obtain two critical points: the first one is the end point of the first-order chiral-transition line turning into the chiral crossover at high temperatures. The second one arises in the low-temperature and high-density region (at least with three degenerate flavors), removing the sharp border line between nuclear matter and superconducting quark matter (this is the so-called quark-hadron continuity). In forthcoming work it will be interesting to investigate further the high-density phase within the nonlocal PNJL model approach.

Concerning the work presented in this work we conclude that the effort of promoting the celebrated NJL model to a nonlocal PNJL model has been worthwhile for several reasons: first the nonlocal Polyakov-loop-extended NJL model can be *deduced* from quantum chromodynamics by performing a suitable transformation and by systematically expanding the resulting theory.

Second, the nonlocal PNJL model meets Dyson-Schwinger calculations (at least in the zero-temperature case), the results of which can be compared to our findings. Third, the nonlocal model does not need to be regularized by a momentum cutoff, unlike the local NJL model, so that the thermodynamics can be described up to arbitrarily large temperatures and densities, in principle. A restriction, in a sense, is imposed upon the model by the degrees of freedom relevant for the description of nuclear matter and its neighborhood. But, there is no conceptual impediment to include diquark or other condensates. Thermal meson modes can be and have been investigated (e. g., Fig. 7.1). Now that lattice computations reach physical pion masses it becomes important to include such mesonic corrections to thermodynamical quantities.

A Notations and Conventions

A.1 Isospin Matrices

The up and down quarks, u and d, form an isospin $SU(2)$ doublet,

$$|u\rangle = \begin{pmatrix} 1 \\ 0 \end{pmatrix}, \qquad |d\rangle = \begin{pmatrix} 0 \\ 1 \end{pmatrix}.$$

The standard isospin basis of the isospin matrices is given by

$$\vec{\tau} = (\tau_1, \tau_2, \tau_3), \qquad \tau_\pm = \frac{1}{2}(\tau_1 \pm i\tau_2)$$

$$\tau_1 = \begin{pmatrix} 0 & 1 \\ 1 & 0 \end{pmatrix}, \qquad \tau_2 = \begin{pmatrix} 0 & -i \\ i & 0 \end{pmatrix}, \qquad \tau_3 = \begin{pmatrix} 1 & 0 \\ 1 & 0 \end{pmatrix},$$

so that

$$\tau_3|u\rangle = |u\rangle, \qquad \tau_3|d\rangle = -|d\rangle, \qquad \tau_-|u\rangle = |d\rangle, \qquad \tau_+ = |u\rangle.$$

The electric charge of a hadron is then given by the Gell-Mann–Nishijima formula

$$Q = I_3 + \frac{1}{2}B,$$

where I_3 is the eigenvalue of the three-component of the isospin operator, $\hat{I}_3 = \frac{1}{2}\tau_3$, and B is the baryon number, $B = \frac{1}{3}$ for quarks and $B = -\frac{1}{3}$ for antiquarks.

Finally, we define

$$\tau_0 = \begin{pmatrix} 1 & 0 \\ 0 & 1 \end{pmatrix},$$

such that the relation

$$\mathrm{tr}\{\tau_i \cdot \tau_j\} = 2\delta_{ij}$$

holds for all $i, j \in \{0, 1, 2, 3\}$.

A.2 Gell-Mann Matrices

The Gell-Mann matrices are one possible representation of the infinitesimal generators of the $SU(3)$ Lie group. This group has dimension eight and therefore it has some set with eight linearly independent generators, $t_a, a \in \{1, \ldots, 8\}$, that obey the commutation relations

$$[t_a, t_b] = i f_{abc} t_c.$$

The structure constants f_{abc} are real and completely *antisymmetric* in the three indices. The nonvanishing structure constants have values

$$f_{123} = 1, \quad f_{147} = f_{165} = f_{246} = f_{257} = f_{345} = f_{376} = \frac{1}{2}, \quad f_{458} = f_{678} = \frac{\sqrt{3}}{2}$$

and fulfill the Jacobi identity

$$f_{abe} f_{ecd} + f_{ace} f_{edb} + f_{ade} f_{ebc} = 0. \tag{A.1}$$

Any set of Hermitian matrices which obey these relations are allowed. Here we state the *fundamental representation* of the SU(3) generators, $t_a := \frac{\lambda_a}{2}$, given in terms of the Gell-Mann matrices $\lambda_a, a \in \{1, \ldots, 8\}$:

$$\lambda_1 = \begin{pmatrix} 0 & 1 & 0 \\ 1 & 0 & 0 \\ 0 & 0 & 0 \end{pmatrix}, \qquad \lambda_2 = \begin{pmatrix} 0 & -i & 0 \\ i & 0 & 0 \\ 0 & 0 & 0 \end{pmatrix}, \qquad \lambda_2 = \begin{pmatrix} 1 & 0 & 0 \\ 0 & -1 & 0 \\ 0 & 0 & 0 \end{pmatrix}$$

$$\lambda_4 = \begin{pmatrix} 0 & 0 & 1 \\ 0 & 0 & 0 \\ 1 & 0 & 0 \end{pmatrix}, \qquad \lambda_5 = \begin{pmatrix} 0 & 0 & -i \\ 0 & 0 & 0 \\ i & 0 & 0 \end{pmatrix}, \qquad \lambda_6 = \begin{pmatrix} 0 & 0 & 0 \\ 0 & 0 & 1 \\ 0 & 1 & 0 \end{pmatrix}$$

$$\lambda_7 = \begin{pmatrix} 0 & 0 & 0 \\ 0 & 0 & -i \\ 0 & i & 0 \end{pmatrix}, \qquad \lambda_8 = \frac{1}{\sqrt{3}} \begin{pmatrix} 1 & 0 & 0 \\ 0 & 1 & 0 \\ 0 & 0 & -2 \end{pmatrix}.$$

These matrices are traceless and Hermitian. In this representation, the product of two Gell-Mann matrices is given by

$$\lambda_a \lambda_b = \frac{2}{N} \delta_{ab} + 2 i f_{abc} \lambda_c + 2 d_{abc} \lambda_c. \tag{A.2}$$

Here we consider $N = 3$, the given relation holds, however, for arbitrary SU(N) Lie algebras in the fundamental representation and the structure constants generalized properly. The d_{abc} are completely *symmetric* coefficients, the nonvanishing values of which are as follows:

$$d_{118} = d_{228} = d_{338} = -d_{888} = \frac{1}{\sqrt{3}}, \quad d_{146} = d_{157} = d_{256} = d_{344} = d_{355} = \frac{1}{2}$$

$$d_{247} = d_{366} = d_{d377} = -\frac{1}{2}, \quad d_{448} = d_{588} = d_{668} = d_{778} = -\frac{1}{2\sqrt{3}}.$$

From Eq. (A.2) it is clear that the d_{abc} are determined by the anticommutator of the Gell-Mann matrices,

$$\{\lambda_a, \lambda_b\} = \frac{4}{N} + 2 d_{abc} \lambda_c.$$

Finally, we define

$$\lambda_0 := \sqrt{\frac{2}{3}} \begin{pmatrix} 1 & 0 & 0 \\ 0 & 1 & 0 \\ 0 & 0 & 1 \end{pmatrix} .$$

Then, the trace relation

$$\mathrm{tr}\{\lambda_\alpha \cdot \lambda_\beta\} = 2\delta_{\alpha\beta}$$

holds for all $\alpha, \beta \in \{0, 1, \ldots, 8\}$, and, thus, generalizes the relation for the isospin (Pauli) matrices.

A.3 Euclidean Dirac Matrices

We define the *Euclidean* Dirac matrices in the chiral representation:

$$\gamma_{\mathrm{E},\mu} = \gamma_{\mathrm{E}}^\mu := (\vec{\gamma}, \gamma_4), \qquad \text{with } \gamma_4 := \mathrm{i}\gamma_0 ,$$

where

$$\gamma_0 = \gamma^0 = \begin{pmatrix} 0 & 1 \\ 1 & 0 \end{pmatrix} , \qquad \vec{\gamma} = \begin{pmatrix} 0 & \vec{\sigma} \\ -\vec{\sigma} & 0 \end{pmatrix}$$

with the 2×2 unit matrix 1 and the Pauli spin matrices

$$\vec{\sigma} = (\sigma_1, \sigma_2, \sigma_3) ,$$

$$\sigma_1 = \begin{pmatrix} 0 & 1 \\ 1 & 0 \end{pmatrix} , \qquad \sigma_2 = \begin{pmatrix} 0 & -\mathrm{i} \\ \mathrm{i} & 0 \end{pmatrix} , \qquad \sigma_3 = \begin{pmatrix} 1 & 0 \\ 1 & 0 \end{pmatrix} .$$

In Euclidean space-time, the matrices fulfill the algebra

$$\{\gamma_\mu, \gamma_\nu\} = -2\delta_{\mu\nu} .$$

Defining the scalar product of two Euclidean four-vectors a, b according to

$$a \cdot b := a_\mu b_\nu \delta_{\mu\nu} := \sum_{i=1}^{4} a_i b_i ,$$

allows one to introduce the "slash" notation

$$\slashed{a} := a_\mu \cdot \gamma_{\mathrm{E}}^\mu = a_\mu \cdot \gamma_{\mathrm{E},\mu} = a_4 \gamma_4 + \vec{a} \cdot \vec{\gamma} .$$

In particular, we have

$$\slashed{\partial} = \gamma_4 \partial_4 + \vec{\gamma} \cdot \nabla ,$$

where $\partial_4 := \frac{\partial}{\partial x_4}$,[1] and x_4 denoting the Euclidean time coordinate introduced in

$$(x_0, \vec{x}\,) \rightarrow (\vec{x}, x_4)\,, \qquad \text{with } x_4 = ix_0\,.$$

---

[1] Note, that this relation is consistent with the definitions $\gamma_4 = i\gamma_0$ and $x_4 = ix_0$, because from the Minkowskian expression, $\partial\!\!\!/ = \gamma^0 \partial_0 + \gamma^i \partial_i$, one obtains with $\gamma^0 \partial_0 = i\gamma^0 \dfrac{\partial}{\partial(ix_0)} = \gamma_4 \dfrac{\partial}{\partial x_4}$ and $\partial_i = \frac{\partial}{\partial x^i} = \nabla$ exactly the Euclidean expression stated previously.

B Derivation of the 't Hooft Interaction

In this appendix we show how the Kobayashi-Maskawa-'t Hooft determinant expression, given in Eq. (3.3.19), can be cast into the form used in this work, Eq. (3.3.21).

In order to write the 't Hooft determinant in a more tractable way, we apply Newton's and Girard's formula

$$\det \mathcal{J}^\pm = \frac{1}{6}\left(\mathrm{tr}\mathcal{J}^\pm\right)^3 - \frac{1}{2}\left(\mathrm{tr}\mathcal{J}^\pm\right)\left(\mathrm{tr}\mathcal{J}^{\pm 2}\right) + \frac{1}{3}\mathrm{tr}\mathcal{J}^{\pm 3}\,. \tag{B.1}$$

Here tr indicates the trace over flavor space only. We use the Gell-Mann matrices as a basis in flavor space, $\{\lambda_0, \lambda_1, \ldots, \lambda_8\}$, with the additional definition $\lambda_0 := \sqrt{\frac{2}{3}}\,\mathrm{diag}(1,1,1)$ in order to maintain $\mathrm{tr}\{\lambda_\alpha \cdot \lambda_\beta\} = 2\delta_{\alpha\beta}$ for all $\alpha, \beta \in \{0, \ldots, 8\}$. This allows one to write

$$\mathcal{J}^\pm = \sum_{\alpha=0}^{8} c_\alpha^\pm \lambda_\alpha \qquad \Longleftrightarrow \qquad \mathrm{tr}\{\lambda_\alpha \mathcal{J}^\pm\} = 2c_\alpha^\pm \tag{B.2}$$

and, consequently, $c_\alpha^\pm = \frac{1}{2}\mathrm{tr}\{\lambda_\alpha \mathcal{J}^\pm\}$.

Furthermore, from Eq. (3.3.19) we have with the definitions (3.3.2)

$$
\begin{aligned}
\frac{1}{2}\mathrm{tr}\{\lambda_\alpha \mathcal{J}^\pm(x)\} &= \frac{1}{4}\int \mathrm{d}^4 z\, \lambda_\alpha^{ij} \bar{\psi}_i\left(x + \frac{z}{2}\right)(1 \mp \gamma_5)\, \mathcal{C}(z)\, \psi_j\left(x - \frac{z}{2}\right) \\
&= \frac{1}{4}j_\alpha^S(x) \mp \frac{1}{4\mathrm{i}}j_\alpha^P(x)\,.
\end{aligned}
$$

By means of Eq. (B.2), this allows one to write $c_\alpha^\pm = \frac{1}{4}j_\alpha^S \pm \frac{1}{4}\mathrm{i}\,j_\alpha^P$, or, inversely $j_\alpha^S = 2\left(c_\alpha^+ + c_\alpha^-\right)$, $j_\alpha^P = -2\,\mathrm{i}\left(c_\alpha^+ - c_\alpha^-\right)$.

Next, we return to Newton's and Girard's formula, Eq. (B.1), and use

$$
\begin{aligned}
\mathrm{tr}\left(\mathcal{J}^\pm\right) &= 2\sqrt{\frac{3}{2}}c_0^\pm \\
\mathrm{tr}\left(\mathcal{J}^{\pm 2}\right) &= \mathrm{tr}\left(c_\alpha^\pm \lambda_\alpha c_\beta^\pm \lambda_\beta\right) = c_\alpha^\pm c_\beta^\pm 2\delta_{\alpha\beta} = 2c_\alpha^\pm c_\alpha^\pm \\
\mathrm{tr}\left(\mathcal{J}^{\pm 3}\right) &= c_\alpha^\pm c_\beta^\pm c_\gamma^\pm \mathrm{tr}\left(\lambda_\alpha \lambda_\beta \lambda_\gamma\right)\,.
\end{aligned}
$$

Inserting this into Eq. (B.1), one has

$$
\begin{aligned}
\det \mathcal{J}^+ + \det \mathcal{J}^- = {} & 2\sqrt{\frac{3}{2}}\left[c_0^{+3} + c_0^{-3}\right] - 2\sqrt{\frac{3}{2}}\left[c_0^+ c_\alpha^+ c_\alpha^+ + c_0^- c_\alpha^- c_\alpha^-\right] + \\
& + \left[c_\alpha^+ c_\beta^+ c_\gamma^+ + c_\alpha^- c_\beta^- c_\gamma^-\right]\mathrm{tr}\left(\lambda_\alpha \lambda_\beta \lambda_\gamma\right)\,.
\end{aligned}
$$

Now, using the relations between the c's and the currents j^S, j^P, we have

$$c_0^{+3} + c_0^{-3} = 2\mathrm{Re}\left[c_0^{+3}\right] = \frac{1}{32}j_0^S\left(j_0^{S2} - 3j_0^{P2}\right)$$

$$c_0^+ c_\alpha^+ c_\alpha^+ + c_0^- c_\alpha^- c_\alpha^- = 2\mathrm{Re}\left[c_0^+ c_\alpha^+ c_\alpha^+\right] = \frac{1}{32}\left[j_0^S\left(j_\alpha^{S2} - j_\alpha^{P2}\right) - 2j_0^P j_\alpha^S j_\alpha^P\right]$$

$$c_\alpha^+ c_\beta^+ c_\gamma^+ + c_\alpha^- c_\beta^- c_\gamma^- = 2\mathrm{Re}\left[c_\alpha^+ c_\beta^+ c_\gamma^+\right] = \frac{1}{32}\left[j_\alpha^S\left(j_\beta^S j_\gamma^S - j_\beta^P j_\gamma^P\right) - j_\alpha^P\left(j_\beta^S j_\gamma^P + j_\gamma^S j_\beta^P\right)\right].$$

Inserting this in the previous formula leads to

$$\det \mathcal{J}^+ + \det \mathcal{J}^- = \frac{1}{16}\sqrt{\frac{3}{2}}j_0^S\left(j_0^{S2} - 3j_0^{P2}\right) - \frac{1}{16}\sqrt{\frac{3}{2}}\left[j_0^S\left(j_\alpha^{S2} - j_\alpha^{P2}\right) - 2j_\alpha^S j_0^P j_\alpha^P\right] +$$
$$+ \frac{1}{96}\left[j_\alpha^S j_\beta^S j_\gamma^S - 3j_\alpha^S j_\beta^P j_\gamma^P\right],$$

where summation over $\alpha, \beta, \gamma \in \{0, \ldots, 8\}$ is implicit.

Finally, using the SU(3) structure constants $f_{k\ell m}, d_{k\ell m}$, defined through $[\lambda_k, \lambda_\ell] = 2\,\mathrm{i}\, f_{k\ell m}\lambda_m$ and $\{\lambda_k, \lambda_\ell\} = \frac{4}{3}\delta_{k\ell} + 2d_{k\ell m}\lambda_m$, respectively, one obtains $\lambda_k \lambda_\ell = \mathrm{i}\, f_{k\ell m}\lambda_m + d_{k\ell m}\lambda_m + \frac{2}{3}\delta_{k\ell}$ and, hence,

$$\mathrm{tr}\left(\lambda_k \lambda_\ell \lambda_i\right) = 2\,\mathrm{i}\, f_{k\ell m}\delta_{mi} + 2d_{k\ell m}\delta_{mi} \tag{B.3}$$

(for $k, \ell, m, i \in \{1, \ldots, 8\}$) which allows one to write

$$\det \mathcal{J}^+ + \det \mathcal{J}^- = \frac{1}{48}\sqrt{\frac{2}{3}}\left(j_0^{S3} - 3j_0^S j_0^{P2}\right) - \frac{1}{32}\sqrt{\frac{2}{3}}\left(j_0^S j_k^S j_k^S - j_0^S j_k^P j_k^P - 2j_k^S j_0^P j_k^P\right) +$$
$$+ \frac{1}{48}d_{\ell k m}\left(j_\ell^S j_k^S j_m^S - 3j_\ell^S j_k^P j_m^P\right),$$

where, again, $\ell, k, m \in \{1, \ldots, 8\}$.

If one sets

$$\mathcal{A}_{\alpha\beta\gamma} := \frac{1}{3!}\varepsilon_{ijk}\varepsilon_{mn\ell}\left(\lambda_\alpha\right)_{im}\left(\lambda_\beta\right)_{jn}\left(\lambda_\gamma\right)_{kl} \qquad \text{for } \alpha, \beta, \gamma \in \{0, \ldots, 8\}, \tag{B.4}$$

then the expression above can be written in a more compact form as

$$\det \mathcal{J}^+ + \det \mathcal{J}^- = \frac{1}{32}\mathcal{A}_{\alpha\beta\gamma}\left(j_\alpha^S j_\beta^S j_\gamma^S - 3j_\alpha^S j_\beta^P j_\gamma^P\right). \tag{B.5}$$

This is the form given in Eq. (3.3.21), with the coupling constant properly adjusted.

C Taylor Expansion of the Euclidean Action $\mathcal{S}_{\mathrm{E}}^{\mathrm{bos}}$

In this appendix we derive the Taylor expansion for the bosonized Euclidean action $\mathcal{S}_{\mathrm{E}}^{\mathrm{bos}}$ up to terms of second order in order to obtain the inverse meson propagators G_P. This corresponds to a systematic perturbative expansion and to the derivation of Feynman rules for the mesonic degrees of freedom according to the method of the "effective action". We would like to point out here, that the bosonized action is composed by two parts: the fermion determinant and a term that is quadratic (for the two-flavor case) or quadratic an cubic (for the three-flavor case) in the Bose fields. While the derivative of the quadratic terms in the two-flavor case is trivial, it is more intricate in the three-flavor case because of the stationary phase approximation, see Eq. (3.3.25). Therefore, we explain the derivative of the fermion part in Sect. C.1 and postpone the treatment of the polynomial terms to Sect. C.2.

C.1 Functional Derivative

We start with a reminder of the Taylor expansion of a functional $T = T[f(x)]$ about $f(0)$ with a (small) fluctuation $h(x)$. One has

$$
\begin{aligned}
T[f(0) + h(x)] = T[f(0)] &+ \int \mathrm{d}y_1 \left. \frac{\delta T[f(x)]}{\delta f(y_1)} \right|_{f=f(0)} h(y_1) + \\
&+ \frac{1}{2} \int \mathrm{d}y_1 \, \mathrm{d}y_2 \left. \frac{\delta^2 T[f(x)]}{\delta f(y_2) \, \delta f(y_1)} \right|_{f=f(0)} h(y_1) h(y_1) + \dots
\end{aligned}
\tag{C.1}
$$

This formalism is now applied to $\mathcal{S}_{\mathrm{E}}^{\mathrm{bos}}$. In this case, we use the function $f(0) = (\{\bar{\sigma}_i\}, \vec{0})$ and the deviation $h(x) = (\{\delta\sigma_i(x)\}, \delta\vec{\pi}(x))$, where $\{\bar{\sigma}_i\}, \{\delta\sigma_i\}$ denote the set of all nonvanishing sigma fields which are just one in the two-, but three ($\sigma_u, \sigma_d, \sigma_s$) in the three-flavor case. The following relation is useful:

$$
\ln \det = \mathrm{Tr} \ln ,
$$

which holds both in functional and matrix space. We treat the functional space first, hence we decompose $\det = \widetilde{\det} \otimes \mathrm{Det}$ or $\mathrm{Tr} = \widetilde{\mathrm{Tr}} \otimes \mathrm{tr}$, where operators with a tilde act exclusively on functional and Det, tr solely on Dirac, flavor and color space.

Then, the zeroth-order term is easy to calculate since $\bar{\sigma}$ is a Lorentz invariant and, hence, proportional to unity in functional space, i.e., $\bar{\sigma}_i \langle p'|p \rangle = \bar{\sigma}_i \frac{(2\pi)^4}{V^{(4)}} \delta^{(4)}(p - p')$. Therefore, the functional trace simply gives an integration over p and the argument of the logarithm is given by $\text{Det}\left[-\not{p}+\hat{m}_q+\mathcal{C}(p)\bar{\sigma}\right]=\prod_{i\in\text{flavors}}(p^2+M_i^2(p))^{2N_c}$. From this the mean-field results (3.2.15) and (3.3.29) are deduced. The linear term vanishes by definition of the mean fields.

In order to determine the term of second order we calculate $\frac{\delta^2}{\delta\sigma_i(k)\,\delta\sigma_j(\ell)}\widetilde{\text{Tr}}\ln\hat{\mathscr{A}}\Big|_{\{\sigma_i(x)\}=\{\bar{\sigma}_i\}}$. As a continuation of matrix multiplication, the logarithm of an operator $\hat{\mathscr{A}}$ is treated as a power series where the multiplication is given by the convolution. Therefore, all matrix identities can be adopted in functional space, too, if all matrix products are replaced by convolutions. In particular, we use operators $\hat{O}=\hat{O}[f]$ that are supposed to fulfill necessary convergence criteria so that the following identities hold:

$$
\begin{aligned}
\frac{\delta}{\delta f(y)}\widetilde{\text{Tr}}\ln\left[\hat{O}[f(x)]\right]&=\widetilde{\text{Tr}}\left[\hat{O}^{-1}[f]\frac{\delta\hat{O}[f(x)]}{\delta f(y)}\right]\\
\frac{\delta}{\delta f(y)}\widetilde{\text{Tr}}\left[\hat{O}^{-1}[f(x)]\right]&=-\widetilde{\text{Tr}}\left[\hat{O}^{-1}[f]\frac{\delta\hat{O}[f(x)]}{\delta f(y)}\hat{O}^{-1}[f]\right].
\end{aligned}
\tag{C.2}
$$

In the case of the operators $\hat{\mathscr{A}}$, Eqs. (3.2.12), (3.3.24), we may write

$$
\frac{\delta^2}{\delta\sigma(k)\,\delta\sigma(\ell)}\widetilde{\text{Tr}}\ln\hat{\mathscr{A}}=-\widetilde{\text{Tr}}\left[(\mathcal{C}\,\delta^{(4)}(\bullet-\ell))\hat{\mathscr{A}}^{-1}(\mathcal{C}\,\delta^{(4)}(\bullet-k))\hat{\mathscr{A}}^{-1}\right],
$$

where the $\bullet$ stands for the arguments of the delta function. Taking into account

$$
\langle p|\hat{\mathscr{A}}|p'\rangle\Big|_{(\{\sigma_i\},\vec{\pi})=(\{\bar{\sigma}_i\},\vec{0})}=(2\pi)^4\delta^{(4)}(p-p')\left[-\not{p}+\hat{m}_q+\mathcal{C}(p)\bar{\sigma}\right],
$$

one carries out the convolutions and arrives at the result:

$$
\begin{aligned}
\frac{\delta^2\ln\det\hat{\mathscr{A}}}{\delta\pi_{k\ell}(\ell)\,\delta\pi_{ij}(k)}&=-\text{Tr}\left\{\hat{\mathscr{A}}^{-1}(p,p')\frac{\delta\hat{\mathscr{A}}}{\delta\pi_{k\ell}}(p',p'')\hat{\mathscr{A}}^{-1}(p'',p''')\frac{\hat{\mathscr{A}}(p''',p)}{\delta\pi_{ij}(k)}\right\}\\
&=-2\text{Tr}\left\{\text{diag}\left(\frac{-\not{p}'+M(p')}{p'^2+M^2(p')}\right)_{jk}\delta(p+p')\,\mathrm{i}\,\gamma_5\sqrt{2}\delta(p'+p''-\ell)\,\mathcal{C}\left(\frac{p'-p''}{2}\right)\right.\\
&\qquad\left.\times\text{diag}\left(\frac{-\not{p}'''+M(p''')}{p'''^2+M^2(p''')}\right)_{\ell i}\delta(p''+p''')\,\mathrm{i}\,\gamma_5\sqrt{2}\delta(p'''+p-k)\,\mathcal{C}\left(\frac{p-p'''}{2}\right)\right\}\\
&=8\,\delta(k+\ell)\delta_{i\ell}\delta_{jk}\int\frac{\mathrm{d}^4p}{(2\pi)^4}\,\mathcal{C}^2\left(p+\frac{k}{2}\right)\frac{p\cdot(p+k)+M_i(p)M_j(p+k)}{\left[p^2+M_i^2(p)\right]\left[(p+k)^2+M_j^2(p+k)\right]}.
\end{aligned}
$$

This is the first part of the second-order term in the Taylor series expansion of the Euclidean action in Eqs. (3.2.13) and (3.3.26). For the two-flavor case, the expansion of the quadratic term in $\{\sigma,\pi_i\}$ of Eq. (3.2.13) is trivially done, leading altogether to the result stated in Eqs. (3.2.16). As mentioned above, the three-flavor case is treated separately in the next section. In Sect. C.3 we investigate the imaginary-pole structure of the loop integrals which becomes important when determining the η' mass.

C.2 Second-Order Contributions to the Action

Let us now consider the polynomial part of the bosonized three-flavor action (3.3.26). Owing to the SPA equations, Eqs. (3.3.25), the auxiliary fields[1] S_α, P_α are implicit functions of σ and π. This implies calculating the second derivative of the expression

$$\tilde{S}_{\rm E} := \sigma_\alpha S_\alpha + \pi_\alpha P_\alpha + \frac{G}{2}\left(S_\alpha S_\alpha + P_\alpha P_\alpha\right) + \frac{H}{4}\mathcal{A}_{\alpha\beta\gamma}\left(S_\alpha S_\beta S_\gamma - 3S_\alpha P_\beta P_\gamma\right).$$

Neglecting first the space dependence of the fields we may first introduce the matrices $\hat{\sigma} = \frac{1}{\sqrt{2}}\sigma_\alpha\lambda_\alpha$ and $\hat{\pi} = \frac{1}{\sqrt{2}}\pi_\alpha\lambda_\alpha$. The SPA equations (3.3.25) in this new basis then read

$$\sqrt{2}\hat{\sigma} + GS_\alpha\lambda_\alpha + \frac{3H}{4}\mathcal{A}_{\alpha\beta\gamma}\lambda_\alpha(S_\beta S_\gamma - P_\beta P_\gamma) = 0 \tag{C.1a}$$

$$\sqrt{2}\hat{\pi} + GP_\alpha\lambda_\alpha - \frac{3H}{2}\mathcal{A}_{\alpha\beta\gamma}\lambda_\alpha S_\beta P_\gamma = 0. \tag{C.1b}$$

From the first derivative of Eq. (C.1a),

$$0 + G\frac{\delta S_\alpha}{\delta\pi_{ij}}\lambda_\alpha + \frac{3H}{4}\mathcal{A}_{\alpha\beta\gamma}\lambda_\alpha\left(2S_\beta\frac{\delta S_\gamma}{\delta\pi_{ij}} - 2P_\beta\frac{\delta P_\gamma}{\delta\pi_{ij}}\right) = 0,$$

it follows that $\frac{\delta S_\alpha}{\delta\pi_{ij}} = 0$ for all $\alpha \in \{0,\dots,8\}$ and $i,j \in \{1,2,3\}$, recalling that $P_\alpha = 0$ for all α in mean-field approximation.
The second derivative of Eq. (C.1a) leads to

$$G\lambda_\alpha\frac{\delta^2 S_\alpha}{\delta\pi_{k\ell}\delta\pi_{ij}} + \frac{3H}{2}\mathcal{A}_{\alpha\beta\gamma}\lambda_\alpha S_\beta\frac{\delta^2 S_\gamma}{\delta\pi_{k\ell}\delta\pi_{ij}} = \frac{3H}{2}\mathcal{A}_{\alpha\beta\gamma}\lambda_\alpha\frac{\delta P_\beta}{\delta\pi_{k\ell}}\frac{\delta P_\gamma}{\delta\pi_{ij}}. \tag{C.2}$$

Analogously, one has from the second equation

$$\sqrt{2}\delta_{im}\delta_{jn} + G\frac{\delta P_\alpha}{\delta\pi_{ij}}(\lambda_\alpha)_{mn} - \frac{3H}{2}\mathcal{A}_{\alpha\beta\gamma}(\lambda_\alpha)_{mn}S_\beta\frac{\delta P_\gamma}{\delta\pi_{ij}} = 0$$

or, by contraction with $(\lambda_\epsilon)_{nm}$

$$G\frac{\delta P_\epsilon}{\delta\pi_{ij}} - \frac{3H}{2}\mathcal{A}_{\epsilon\beta\gamma}S_\beta\frac{\delta P_\gamma}{\delta\pi_{ij}} = -\frac{1}{\sqrt{2}}(\lambda_\epsilon)_{ij}. \tag{C.3}$$

Finally, from the second derivative

$$0 + G\frac{\delta^2 P_\alpha}{\delta\pi_{k\ell}\delta\pi_{ij}}(\lambda_\alpha)_{mn} - \frac{3H}{2}\mathcal{A}_{\alpha\beta\gamma}(\lambda_\alpha)_{mn}S_\beta\frac{\delta^2 P_\gamma}{\delta\pi_{k\ell}\delta\pi_{ij}} = 0,$$

and it follows that $\frac{\delta^2 P_\alpha}{\delta\pi_{k\ell}\delta\pi_{ij}} = 0$ for all α, i, j, k, ℓ in mean-field approximation.
The sum of the SPA equations gives

$$\sigma_\alpha S_\alpha + \pi_\alpha P_\alpha + G(S_\alpha S_\alpha + P_\alpha P_\alpha) + \frac{3H}{4}\mathcal{A}_{\alpha\beta\gamma}(S_\alpha S_\beta S_\gamma - 3S_\alpha P_\beta P_\gamma) = 0,$$

---

[1] From now on we omit the tildes on $\tilde{S}_\alpha, \tilde{P}_\alpha$.

so that one can write

$$\tilde{\mathcal{S}}_{\mathrm{E}} = -\frac{1}{2}G(S_\alpha S_\alpha + P_\alpha P_\alpha) - \frac{H}{2}\mathcal{A}_{\alpha\beta\gamma}(S_\alpha S_\beta S_\gamma - 3S_\alpha P_\beta P_\gamma)\,.$$

Finally, applying identities (C.2) and (C.3) we may deduce the desired derivative

$$\frac{\delta^2\tilde{\mathcal{S}}_{\mathrm{E}}}{\delta\pi_{k\ell}\,\delta\pi_{ij}} = \frac{1}{\sqrt{2}}\,(\lambda_\beta)_{ij}\,\frac{\delta P_\beta}{\delta\pi_{k\ell}}\,.$$

We conclude that the additional term is given by the solution of relation (C.3) contracted by λ_α,

$$G\frac{\delta P_\alpha}{\delta\pi_{ij}}(\lambda_\alpha)_{mn} - \frac{3H}{2}\mathcal{A}_{\alpha\beta\gamma}(\lambda_\alpha)_{mn}S_\beta\frac{\delta P_\gamma}{\delta\pi_{ij}} = -\sqrt{2}\delta_{im}\delta_{jn}\,;$$

this can be further simplified by noting $S_\alpha = \frac{1}{2}\mathrm{tr}(\lambda_\alpha S)$ and

$$\begin{aligned}
\mathcal{A}_{\alpha\beta\gamma}(\lambda_\alpha)_{mn}S_\beta\frac{\delta P_\gamma}{\delta\pi_{ij}} &= \frac{1}{3!}\varepsilon_{rsk}\varepsilon_{uv\ell}(\lambda_\alpha)_{ru}(\lambda_\beta)_{sv}(\lambda_\gamma)_{k\ell}\frac{1}{2}S_t(\lambda_\beta)_{tt}(\lambda_\alpha)_{mn}\frac{\delta P_\gamma}{\delta\pi_{ij}} \\
&= \frac{1}{3}\varepsilon_{ntk}\varepsilon_{mt\ell}\frac{\delta P_\gamma}{\delta\pi_{ij}}(\lambda_\gamma)_{k\ell}S_t\,.
\end{aligned}$$

Consequently, the equation to be solved is

$$G\,(\lambda_\alpha)_{mn}\frac{\delta P_\alpha}{\delta\pi_{ij}} - \frac{H}{2}\varepsilon_{knt}\varepsilon_{t\ell m}S_t\,(\lambda_\gamma)_{k\ell}\frac{\delta P_\gamma}{\delta\pi_{ij}} = -\sqrt{2}\delta_{im}\delta_{jn}\,.$$

Defining $(r_{ij,mn})^{-1} := \frac{1}{\sqrt{2}}(\lambda_\alpha)_{mn}\frac{\delta P_\alpha}{\delta\pi_{ij}}$ we may write

$$\frac{\delta^2\tilde{\mathcal{S}}_{\mathrm{E}}}{\delta\pi_{k\ell}\,\delta\pi_{ij}} = -(r_{ij,k\ell})^{-1}\,, \tag{C.4}$$

where $r_{ij,k\ell}$ solves the system given in Eq. (3.3.37).

Finally, we consider the functional derivative of terms of the form $\int \mathrm{d}^4x\, S_\alpha(x)S_\beta(x)S_\gamma(x)$ etc. The first derivative with respect to $\pi_{ij}(y)$ generates a delta function, $\delta(x-y)$, hence

$$\int \mathrm{d}^4x\, S_\alpha(x)S_\beta(x)S_\gamma(x) \to S_\alpha(y)S_\beta(y)S_\gamma(y)\,.$$

The second derivative with respect to $\pi_{k\ell}(z)$ generates an additional $\delta(y-z)$. This means that in mean-field approximation the functional dependence of the fields after a Fourier transformation is given by

$$r_{ij,k\ell}^{-1}\int \mathrm{d}^4y\,\mathrm{d}^4z\, \mathrm{e}^{-\mathrm{i}p\cdot y}\,\mathrm{e}^{-\mathrm{i}p'\cdot z}\,\delta(y-z)\,\delta\pi_{ij}(y)\,\delta\pi_{k\ell}(z) = r_{ij,k\ell}^{-1}\,\delta\pi_{ij}(p)\,\delta\pi_{k\ell}(-p)\,.$$

Treating analogously the contributions from the σ field, and combining this with the derivatives of the fermion determinant, calculated in the previous section, we arrive at Eq. (3.3.34).

C.3 Evaluation of Quark Loop Integrals

When determining formulas for the meson masses in Sect. 3.3.3 we have mentioned that one of the advantages of the nonlocal NJL model is that the integrand in Eq. (3.3.36) might not have poles on the imaginary axis. Since all formulas have been derived in Euclidean space, this means that the mesons do not decay (unphysically) into a pair of constituent quarks. Whether or not such poles arise is, however, dependent on the form of the distribution function $\mathcal{C}(p)$ and the explicit values of the input parameters. For the choice of the parameters used in this work it turns out that only the integrals, Eq. (3.3.36), involved in the determination of the η' mass are affected by poles on the imaginary plane. Here we show how these integrals have to be treated properly.

Suppose that the meson mass is given by p_0. Then, without loss of generality, one can choose $p = (\vec{0}, ip_0)$ and use three-dimensional rotational invariance to write the quark loop integrals determining the meson mass as

$$\tilde{\Pi}_{ij}(-p_0^2) = \int \mathrm{d}q_3 \, \mathrm{d}q_4 \, q_3^2 \frac{F_{ij}(q_3, q_4, p_0)}{\left[q^{+2} + M_i^2(q^+)\right]\left[q^{-2} + M_j^2(q^-)\right]}, \qquad (C.1)$$

where $q_3 := |\vec{q}|$ and $q^\pm = \left(0, 0, q_3, q_4 \pm i\frac{p_0}{2}\right)$. The explicit form of $F_{ij}(q_3, q_4, p_0)$ depends on the described meson and can be read off by comparing Eqs. (C.1) and (3.3.36). As mentioned previously, the integral (C.1) has poles in the integration domain $(q_3, q_4) \in [0, \infty) \times (-\infty, \infty)$— and is hence divergent—only if $p_0 = im_{\eta'}$. For this case, only expressions $\tilde{\Pi}_{ij}$ with $i = j$ are relevant, see Eqs. (3.3.47) and (3.3.40), (3.3.41), (3.3.42). From now on, we omit the indices i, j. Let us first determine the zeros of the denominator in Eq. (C.1). We write the denominator as $D = D^+ D^-$ with

$$D^\pm = \left(q^\pm\right)^2 + M^2(q^\pm). \qquad (C.2)$$

In order to find the zeros of D, we first characterize the zeros ξ_ν of the inverse quark propagator $S^{-1}(q) = -q\!\!\!/ + M(q)$ by a set of two real numbers (S_r^ν, S_i^ν), with $S_r^\nu \in \mathbb{R}_0^+$ and $S_i^\nu \in \mathbb{R}^+$. Since the zeros appear in multiplets, we have introduced the index $\nu \in \mathbb{N}$ that labels these multiplets. It turned out, however, that for the particular choice of the distribution function $\mathcal{C}(q)$ chosen in this work only one multiplet of zeros appears, therefore the index ν will be dropped in the following.[2] Furthermore, one would have to distinguish, in principle, between two cases: the first, in which purely imaginary poles exist, i.e., $S_r = 0$, and the second, with no purely imaginary poles, i.e., $S_r > 0$. In our case, only the latter is important. It turns out, then, that such complex poles appear as quartets located at $\sqrt{\xi^2} = S_r \pm iS_i$ and $\sqrt{\xi^2} = -S_r \pm iS_i$. Having determined the zeros of S^{-1}, we find the zeros of D—or, equivalently the zeros of D^+ and/or D^-—as

$$q^{+2} = q_3^2 + q_4^2 - \frac{p_0^2}{4} + iq_4 p_0 \equiv S_r^2 - S_i^2 \pm 2iS_r S_i \qquad (C.3a)$$

---

[2]Even if there were more poles appearing, the effect of these higher poles would not be observed as long as p_0 is considerably lower than the S_i of the corresponding pole.

and/or

$$q^{-2} = q_3^2 + q_4^2 - \frac{p_0^2}{4} - iq_4 p_0 \equiv S_r^2 - S_i^2 \pm 2iS_r S_i . \tag{C.3b}$$

Solving these equations for q_4 leads to eight different solutions; the zeros of D^+ are

$$q_4^{(3,1)} = -\frac{S_i S_r}{\gamma(q_3, S_i, S_r)} + i\left(\pm\gamma(q_3, S_i, S_r) - \frac{p_0}{2}\right) \tag{C.4a}$$

and the poles of D^- are

$$q_4^{(4,2)} = -\frac{S_i S_r}{\gamma(q_3, S_i, S_r)} + i\left(\pm\gamma(q_3, S_i, S_r) + \frac{p_0}{2}\right) , \tag{C.4b}$$

with

$$\gamma(q_3, S_i, S_r) = \sqrt{\frac{q_3^2 + (S_i^2 - S_r^2) + \sqrt{q_3^4 + 2q_3^2(S_i^2 - S_r^2) + (S_i^2 + S_r^2)^2}}{2}} , \tag{C.5}$$

and the other four poles are given by

$$q_4^{(4+i)} = -\mathrm{Re}\left(q_4^{(i)}\right) + i\,\mathrm{Im}\left(q_4^{(i)}\right) \qquad \text{for } i \in \{1, \dots, 4\} . \tag{C.6}$$

Clearly, for small values of p_0, q_4 cannot be real (since S^{-1} does not have real zeros as $p_0 \to 0$). Now, as p_0 increases, the poles move closer towards the real axis until they meet on the real q_4 axis at the point (q_3^p, q_4^p). This is fixed by the condition $\mathrm{Im}\,q_4^{(2)} = \mathrm{Im}\,q_4^{(3)} = 0$, hence, using Eqs. (C.4b) and (C.4a)

$$(q_3^p, q_4^p) = \left(\frac{\sqrt{(p_0^2 - 4S_i^2)(p_0^2 + 4S_r^2)}}{2p_0}, \pm\frac{2S_i S_r}{p_0}\right) . \tag{C.7}$$

From this, we clearly see, that such a point only exists for $q_3 \geq 2S_i$.[3]

Having isolated the pole in the integration region, we can now describe the proper regularization procedure. This requires to introduce two small parameters ε and δ and to take the limit $\delta \to 0^+, \varepsilon \to 0^+$ at the end of the calculation. ε (corresponding to the Feynman ε) is used to shift the poles of D^+ and D^-, whereas δ is used to split the q_3 integration interval in three subintervals: $q_3 > q_3^p + \delta, q_3^p - \delta < q_3 < q_3^p + \delta$ and $q_3 < q_3^p - \delta$. In the regions $q_3 > q_3^p + \delta$ and $q_3 < q_3^p - \delta$ the integration over the real q_4 axis can be performed, and the limit $\varepsilon \to 0^+$ can be taken even before performing the integrations. In the region $q_3^p - \delta < q_3 < q_3^p + \delta$ the situation is, however, more complicated, because the integration over q_4 hits the poles at $q_4 = q_4^p$ for $q_3 = q_3^p$. Therefore, the q_4 integration over $(-\infty, \infty)$ is shifted to the contour $q_4 \to q_4 + i\kappa\delta$, with $\kappa > 1$ being an arbitrary constant. The final result can then be obtained by applying the residue theorem to the complex integrals described above.

In the case $S_r \neq 0$ primarily considered here, one has an ambiguity in the choice of the sign

---

[3]In the case $S_r \neq 0$ there exist two of them, as it is evident from Eq. (C.7).

of the (Euclidean) ε prescription. Choosing different signs of ε for sets $q_4^{(i)}, i \in \{1, \ldots, 4\}$ and $q_4^{(j)}, j \in \{5, \ldots, 8\}$ (see Ref. [61]), it is not hard to see that the contributions to the imaginary part of the quark loop integral vanish:

$$\mathrm{Im}\left[\tilde{\Pi}(-p_0^2)\right] = 0 . \tag{C.8a}$$

The real part is given by

$$\mathrm{Re}\left[\tilde{\Pi}(-p_0^2)\right] = \lim_{\delta \to 0^+} \left\{ R(-p_0^2, \delta) + \int_0^{q_3^p - \delta} dq_3 \int_{-\infty}^{\infty} dq_4 \frac{q_3^2 F(q_3, q_4, p_0)}{\left[q^{+2} + M^2(q^+)\right]\left[q^{-2} + M^2(q^-)\right]} + \right.$$

$$\left. + \int_{q_3^p + \delta}^{\infty} dq_3 \int_{-\infty}^{\infty} dq_4 \frac{q_3^2 F(q_3, q_4, p_0)}{\left[q^{+2} + M^2(q^+)\right]\left[q^{-2} + M^2(q^-)\right]} \right\} . \tag{C.8b}$$

Here $R(-p_0^2, \delta)$ is the so-called "residue contribution", responsible for the cancelation of the divergence appearing in the integrals in Eq. (C.8b) in the limit $\delta \to 0^+$. It follows from the residue theorem as

$$R(-p_0^2, \delta) = 4\pi \int_0^{q_3^p - \delta} dq_3 \,\mathrm{Re}\left[\frac{q_3^2 F(q_3, q_4, p_0)}{\left[q^{+2} + M^2(q^+)\right]\left[1 + \partial M^2(q^-)/\partial(q^-)^2\right](iq_4 + p_0/2)} \right]_{q_4 = q_4^{(2)}} .$$

For the sake of completeness, we also state the result for the case $S_r = 0$ (see Ref. [62]). For the imaginary part (which is solely determined by the integration region $q_3^p - \delta < q_3 < q_3^p + \delta$) we obtain

$$\mathrm{Im}\left[\tilde{\Pi}(-p_0^2)\right] = -\frac{\pi^2}{2p_0} \frac{q_3^p F(q_3^p, 0, p_0)}{\left[1 + \frac{\partial M^2(q)}{\partial q^2}\Big|_{q^2 = -S_i^2}\right]^2} \tag{C.9a}$$

while the real part is the same as in Eq. (C.8b) with the residue contribution replaced by the following expression:

$$\mathrm{Re}\left[\tilde{\Pi}(-p_0^2)\right] = \text{Eq. (C.8b) with}$$

$$R(-p_0^2, \delta) = 2\pi \int_0^{q_3^p - \delta} \frac{dq_3}{\sqrt{q_3^2 + S_i^2}} \,\mathrm{Re}\left[\frac{q_3^2 F(q_3, q_4, p_0)}{\left[q^{+2} + M^2(q^+)\right]\left[1 + \partial M^2(q^-)/\partial(q^-)^2\right]} \right]_{q_4 = q_4^{(2)}(S_r = 0)} . \tag{C.9b}$$

Using the expressions Eqs. (C.8) allows one to determine the η' mass according to Eq. (3.3.43).

D Derivation of the Pseudoscalar Meson Decay Constants

In this appendix section we present a detailed derivation of the formulas for the (pseudoscalar) decay constants[1] which were defined as

$$\langle 0|J_{A,\alpha}^{\mu}(0)|\tilde{\phi}_{\beta}(p)\rangle = \mathrm{i}\, f_{\alpha\beta}\, p_{\mu} \quad \Longleftrightarrow \quad \langle 0|J_{A,\alpha}^{\mu}(0)|\phi_{\beta}(p)\rangle = \mathrm{i}\, f_{\alpha\beta} Z_{\phi}^{1/2}\, p_{\mu}, \tag{3.3.49}$$

where $J_{A,\alpha}^{\mu}$ denotes the axial-vector current. In the following we outline the calculation of the (unrenormalized) matrix element $\langle 0|J_{A,\alpha}^{\mu}(0)|\phi_{\beta}(p)\rangle$.

In order to calculate the matrix element one has to gauge the nonlocal action in Eq. (3.3.22). As discussed in Sects. 3.2.2 and 3.3.3, gauge invariance requires not only the replacement of the partial derivative by a covariant derivative,

$$\partial_{\mu} \to \partial_{\mu} + \frac{\mathrm{i}}{2}\gamma_{5}\lambda_{\alpha}\,\mathcal{A}_{\mu}^{\alpha}(x)\,,$$

where $\mathcal{A}_{\mu}^{\alpha}$ ($\alpha \in \{0\ldots,8\}$) are a set of axial gauge fields, but also the connection of nonlocal terms through a parallel transport with a Wilson line,

$$\mathcal{W}(x,y) = \mathcal{P}\exp\left\{\frac{\mathrm{i}}{2}\int_{0}^{1}\mathrm{d}\alpha\,\gamma_{5}\lambda_{\alpha}\,\mathcal{A}_{\alpha}^{\mu}(x+(y-x)\alpha)\,(y_{\mu}-x_{\mu})\right\}\,,$$

where we have chosen a straight line that connects the points x and y. Introducing the auxiliary fields in the bosonization procedure properly, one gets an action in which only the fermion determinant $\mathscr{A}$ is affected by the gauging. In coordinate space this reads

$$\begin{aligned}
\mathscr{A}^{\mathrm{G}}(x,y) &= \left(-\mathrm{i}\,\partial\!\!\!/_{y} + \frac{1}{2}\gamma_{5}\lambda_{\alpha}\,\mathcal{A}^{\alpha} + \hat{m}_{c}\right)\delta(x-y)+ \\
&+ \mathcal{C}(x-y)\,\mathcal{W}\left(x,\frac{x+y}{2}\right)\Gamma_{\alpha}\,S_{\alpha}\left(\frac{x+y}{2}\right)\mathcal{W}\left(\frac{x+y}{2},y\right),
\end{aligned} \tag{3.3.50}$$

where Γ_{α} stands either for $\Gamma_{\alpha}=\lambda_{\alpha}$ or $\Gamma_{\alpha}=\mathrm{i}\gamma_{5}\lambda_{\alpha}$, and S_{α} accordingly for either a scalar field, σ_{α}, or a pseudoscalar field, π_{α}.

The desired matrix element then follows from the gauged fermion determinant according to

$$\langle 0|J_{A,\alpha}^{\mu}(0)|\pi_{\beta}(p)\rangle = -\left.\frac{\delta^{2}\ln\det\mathscr{A}^{\mathrm{G}}}{\delta\pi_{\beta}(p)\,\delta\mathcal{A}_{\mu}^{\alpha}(t)}\right|_{\substack{A=0\\t=0}}. \tag{D.1}$$

---

[1]We treat here the three-flavor case, because the two-flavor result is contained in this more general result, as mentioned in the text.

We begin with the calculation of the expressions needed, determining the complete matrix element. One has

$$\left.\frac{\delta \mathcal{W}(x,y)}{\delta \mathcal{A}^{\nu}_{\beta}(t)}\right|_{\substack{\mathcal{A}=0 \\ t=0}} = \frac{i}{2}\int_0^1 \mathrm{d}\alpha\, \gamma_5 \lambda_\beta\, \delta(x+(y-x)\alpha)\,(y_\nu - x_\nu)$$

and using Eq. (C.2), properly modified

$$\left.\frac{\delta \ln \det \mathscr{A}^{G}(x,y)}{\delta \mathcal{A}^{\nu}_{\beta}}\right|_{\substack{\mathcal{A}=0 \\ t=0}} = -\mathrm{Tr}\Bigg\{\Bigg[\delta(x-y)\frac{1}{2}\gamma_5\lambda_\beta\gamma^{\nu}\,\delta(y)$$

$$+\, C(x-y)\frac{i}{2}\int_0^1 \mathrm{d}\alpha\,\gamma_5\lambda_\beta\,\delta\Big(x+\Big(\frac{x+y}{2}-x\Big)\alpha\Big)\Big(\frac{x_\nu+y_\nu}{2}-x_\nu\Big)\Gamma_\alpha S_\alpha\Big(\frac{x+y}{2}\Big)$$

$$-\, C(x-y)\frac{i}{2}\,\Gamma_\alpha S_\alpha\Big(\frac{x+y}{2}\Big)\int_0^1 \mathrm{d}\alpha\,\gamma_5\lambda_\beta\,\delta\Big(y+\Big(\frac{x+y}{2}-y\Big)\alpha\Big)\Big(\frac{x_\nu+y_\nu}{2}-y_\nu\Big)\Bigg]$$

$$\times\,\mathscr{A}^{-1}(y,y')\Bigg\}.$$

We need now to calculate the derivative of this expression with respect to $\pi_\beta(p)$, i.e., with respect to a momentum-dependent variable. Owing to

$$\tilde{\mathrm{Tr}}\,[A(x,\xi)B(\xi,y)] = \tilde{\mathrm{Tr}}\,\Big[\langle x|\hat{A}|\xi\rangle\langle\xi|\hat{B}|x\rangle\Big]$$

$$= \tilde{\mathrm{Tr}}\,\Big[\langle x|p\rangle\langle p|\hat{A}|p'\rangle\langle p'|\xi\rangle\langle\xi|k\rangle\langle k|B|k'\rangle\langle k'|x\rangle\Big]$$

$$= \tilde{\mathrm{Tr}}\,\Big[\langle p|\hat{A}|p'\rangle\langle p'|\hat{B}|p\rangle\Big]$$

$$= \tilde{\mathrm{Tr}}\,[A(p,p')B(p',q)]$$

the functional trace of a product of operators can be calculated in an arbitrary basis as the product of the operators, each of them represented in that basis. Since we know already the momentum representation of $\hat{\mathscr{A}}^{-1}$, it is now sufficient to calculate the Fourier transforms of the expressions in the squared bracket above. We have

$$\int \mathrm{d}^4x\,\mathrm{d}^4y\, e^{-i\,q\cdot x}\, e^{-i\,q'\cdot y}\,\delta(x-y)\frac{1}{2}\gamma_5\lambda_\beta\gamma^{\nu}\,\delta(y) = \frac{1}{2}\gamma_5\lambda_\beta\gamma^{\nu},$$

and, using in the following the substitution $u = x - y, v = \frac{x+y}{2} \Leftrightarrow x = v + \frac{u}{2}, y = v - \frac{u}{2},$

$$\frac{i}{2}\int \mathrm{d}^4x\,\mathrm{d}^4y\, e^{-i\,q\cdot x}\, e^{-i\,q'\cdot y}\, C(x-y)\int_0^1 \mathrm{d}\alpha\,\gamma_5\lambda_\beta\,\delta\Big(x+\Big(\frac{x+y}{2}-x\Big)\alpha\Big)\Big(\frac{x_\nu+y_\nu}{2}-x_\nu\Big)\Gamma_\alpha S_\alpha\Big(\frac{x+y}{2}\Big)$$

$$= \frac{i}{2}\int \mathrm{d}^4u\,\mathrm{d}^4v\int_0^1 \mathrm{d}\alpha\, e^{-i\,q\cdot(v+\frac{u}{2})}\, e^{-i\,q'\cdot(v-\frac{u}{2})}\, C(u)\gamma_5\lambda_\beta\,\delta\Big(v+\frac{u}{2}-\frac{u}{2}\alpha\Big)\Big(-\frac{u_\nu}{2}\Big)\Gamma_\alpha S_\alpha(v)$$

$$= -\frac{i}{2}\int \mathrm{d}^4u\int_0^1 \mathrm{d}\alpha\, e^{-i\,q\cdot\frac{u}{2}\alpha}\, e^{-i\,q'\cdot(-u+\frac{u}{2}\alpha)}\,\frac{u_\nu}{2}\, C(u)\gamma_5\lambda_\beta\,\Gamma_\alpha S_\alpha\Big(-\frac{u}{2}(1-\alpha)\Big)$$

$$= -\frac{i}{4}\int \mathrm{d}^4u\,\frac{\mathrm{d}^4k}{(2\pi)^4}\int_0^1 \mathrm{d}\alpha\, e^{-i\,q\cdot\frac{u}{2}\alpha}\, e^{-i\,q'\cdot u(-1+\frac{\alpha}{2})}\, u_\nu C(u)\,\gamma_5\lambda_\beta\,\Gamma_\alpha\, e^{-i\,k\cdot\frac{u}{2}(1-\alpha)}\, S_\alpha(k).$$

Analogously we have

$$-\frac{i}{2}\int d^4x\, d^4y\, e^{-i\,q\cdot x}\, e^{-i\,q'\cdot y}\, \mathcal{C}(x-y)\int_0^1 d\alpha\, \Gamma_\alpha S_\alpha\!\left(\frac{x+y}{2}\right)\gamma_5\lambda_\beta\, \delta\!\left(y+\left(\frac{x+y}{2}-y\right)\alpha\right)\!\left(\frac{x_\nu+y_\nu}{2}-y_\nu\right)$$

$$=+\frac{i}{4}\int d^4u\, \frac{d^4k}{(2\pi)^4}\int_0^1 d\alpha\, e^{-i\,q'\cdot\frac{u}{2}\alpha}\, e^{-i\,q\cdot u\left(-1+\frac{\alpha}{2}\right)}\, u_\nu \mathcal{C}(u)\,\Gamma_\alpha \gamma_5\lambda_\beta\, e^{-i\,k\cdot\frac{u}{2}(1-\alpha)}\, S_\alpha(k).$$

Having this, we can write down the desired expression in mean-field approximation ($\Gamma_\alpha S_\alpha(k)\to$ $\mathrm{diag}(\bar\sigma_i)\,\delta(k)$)

$$\frac{\delta^2 \ln\det \mathscr{A}^G}{\delta\pi_\gamma(\ell)\,\delta A_\beta^\nu(t)}\bigg|_{\substack{A=0\\ t=0}} = -\mathrm{Tr}\left\{\left[-\frac{i}{4}\int d^4u\int_0^1 d\alpha\, e^{-i\,q\cdot\frac{u}{2}\alpha}\, e^{-i\,q'\cdot u\left(-1+\frac{\alpha}{2}\right)}\, u_\nu \mathcal{C}(u)\,\gamma_5\lambda_\beta\,\Gamma_\gamma\, e^{-i\,\ell\cdot\frac{u}{2}(1-\alpha)}+\right.\right.$$

$$+\frac{i}{4}\int d^4u\int_0^1 d\alpha\, e^{-i\,q'\cdot\frac{u}{2}\alpha}\, e^{-i\,q\cdot u\left(-1+\frac{\alpha}{2}\right)}\, u_\nu \mathcal{C}(u)\,\Gamma_\gamma\, \gamma_5\lambda_\beta\, e^{-i\,\ell\cdot\frac{u}{2}(1-\alpha)}\bigg]\times$$

$$\times \mathrm{diag}\!\left(\frac{-q''+M(q'')}{q''^2+M^2(q'')}\right)\delta(q''+q')\delta(q-q'')\bigg\}+$$

$$+\,\mathrm{Tr}\left\{\left[\frac{1}{2}\gamma_5\lambda_\beta\gamma^\nu-\frac{i}{4}\int d^4u\int_0^1 d\alpha\, e^{-i\,q\cdot\frac{u}{2}\alpha-i\,q'\cdot u\left(-1+\frac{\alpha}{2}\right)}\, u_\nu \mathcal{C}(u)\gamma_5\lambda_\beta\,\mathrm{diag}(\bar\sigma_i)+\right.\right.$$

$$+\frac{i}{4}\int d^4u\int_0^1 d\alpha\, e^{-i\,q'\cdot\frac{u}{2}\alpha-i\,q\cdot u\left(-1+\frac{\alpha}{2}\right)}\, u_\nu \mathcal{C}(u)\,\mathrm{diag}(\bar\sigma_i)\,\gamma_5\lambda_\beta\bigg]\times$$

$$\times\,\mathrm{diag}\!\left(\frac{-q''+M(q'')}{q''^2+M^2(q'')}\right)\delta(q'+q'')\mathcal{C}\!\left(\frac{q''-q'''}{2}\right)\Gamma_\gamma\delta(q''+q'''-\ell)\times$$

$$\times\mathrm{diag}\!\left(\frac{-q+M(q)}{q^2+M^2(q)}\right)\delta(q'''+q)\bigg\}.$$

Now, we show how the above expression can be simplified term by term. We start with the first summand, which after performing the Fourier transform, gives (note, that there are contributions only from $\Gamma_\gamma=i\gamma_5\lambda_\gamma)^2$

$$\frac{1}{4}\mathrm{Tr}\left\{\int_0^1 d\alpha\left[\left(2s_\nu\frac{d\mathcal{C}}{ds^2}\right)_{\frac{q\alpha}{2}+q'\left(-1+\frac{\alpha}{2}\right)+\frac{\ell}{2}(1-\alpha)}\gamma_5\lambda_\beta\Gamma_\gamma-\left(2s_\nu\frac{d\mathcal{C}}{ds^2}\right)_{\frac{q'\alpha}{2}+q\left(-1+\frac{\alpha}{2}\right)+\frac{\ell}{2}(1-\alpha)}\Gamma_\gamma\gamma_5\lambda_\beta\right]\times\right.$$

$$\times\mathrm{diag}\!\left(\frac{-q+M(q)}{q^2+M^2(q)}\right)\delta(q+q')\bigg\}$$

$$=\frac{1}{2}\mathrm{Tr}\left\{\int_0^1 d\alpha\left[\left(s_\nu\frac{d\mathcal{C}}{ds^2}\right)_{q+\frac{\ell}{2}(1-\alpha)}\gamma_5\lambda_\beta\Gamma_\gamma-\left(s_\nu\frac{d\mathcal{C}}{ds^2}\right)_{-q+\frac{\ell}{2}(1-\alpha)}\Gamma_\gamma\gamma_5\lambda_\beta\right]\times\mathrm{diag}\!\left(\frac{-q+M(q)}{q^2+M^2(q)}\right)\right\}$$

$$=2i\,\mathrm{tr}\left\{\int_0^1 d\alpha\left(q_\nu+\frac{\ell_\nu}{2}(1-\alpha)\right)\frac{d\mathcal{C}}{ds^2}\bigg|_{q+\frac{\ell}{2}(1-\alpha)}\lambda_\beta^{ij}\lambda_\gamma^{ji}\frac{M_i(q)}{q^2+M_i^2(q)}+\right.$$

$$+\int_0^1 d\alpha\left(q_\nu-\frac{\ell_\nu}{2}(1-\alpha)\right)\frac{d\mathcal{C}}{ds^2}\bigg|_{-q+\frac{\ell}{2}(1-\alpha)}\lambda_\gamma^{ij}\lambda_\beta^{ji}\frac{M_i(q)}{q^2+M_i^2(q)}\bigg\}$$

$$=2i\left(\lambda_\beta^{ij}\lambda_\gamma^{ji}+\lambda_\gamma^{ij}\lambda_\beta^{ji}\right)\mathrm{tr}\left\{\int_0^1 d\alpha\, q_\nu\frac{d\mathcal{C}}{dq^2}\frac{M_i\!\left(q+\frac{\ell}{2}(1-\alpha)\right)}{\left(q+\frac{\ell}{2}(1-\alpha)\right)^2+M_i^2\!\left(q+\frac{\ell}{2}(1-\alpha)\right)}\right\}.$$

---

[2] Tr denotes the traces over Dirac, flavor and color indices as well as the functional trace, while tr stands only for the functional trace.

The first summand in the second term leads

$$
- \mathrm{Tr} \left\{ \frac{1}{2} \gamma_5 \lambda_\beta^{ij} \gamma^\nu \frac{-\slashed{q}' + M_j(q')}{q'^2 + M_j^2(q')} \, \mathcal{C}\left(q' + \frac{\ell}{2}\right) \mathrm{i}\, \gamma_5 \lambda_\gamma^{ji} \frac{-(\slashed{q}' + \slashed{\ell}) + M_i(q' + \ell)}{(q'+\ell)^2 + M_i^2(q'+\ell)} \right\}
$$

$$
= -2\mathrm{i}\, \lambda_\beta^{ij} \lambda_\gamma^{ji} \, \mathrm{tr} \left\{ \frac{-q^\nu M_i(q+\ell) + (q^\nu + \ell^\nu) M_j(q)}{\left(q^2 + M_j^2(q)\right)\left((q+\ell)^2 + M_i^2(q+\ell)\right)} \mathcal{C}\left(q + \frac{\ell}{2}\right) \right\}
$$

$$
= 2\mathrm{i} \lambda_\beta^{ij} \lambda_\gamma^{ji} \, \mathrm{tr} \left\{ \frac{\mathcal{C}(q)\left(q^\nu + \frac{\ell^\nu}{2}\right) M_i\left(q - \frac{\ell}{2}\right)}{\left(q^{+2} + M_j^2(q^+)\right)\left(q^{-2} + M_i^2(q^-)\right)} + \frac{(q^\nu - \ell^\nu) M_j(q)}{\left(q^2 + M_j^2(q)\right)((q-\ell)^2 + M_i^2(q-\ell))} \mathcal{C}\left(q - \frac{\ell}{2}\right) \right\}
$$

$$
= 2\mathrm{i} \left(\lambda_\beta^{ij} \lambda_\gamma^{ji} + \lambda_\beta^{ji} \lambda_\gamma^{ij} \right) \mathrm{tr} \left\{ \mathcal{C}(q) \frac{q^\nu{}^+ M_i(q^-)}{\left(q^{+2} + M_j^2(q^+)\right)\left(q^{-2} + M_i^2(q^-)\right)} \right\}.
$$

The second summand of the second term gives

$$
- \frac{1}{4} \mathrm{Tr} \left\{ \int_0^1 d\alpha \left(2 s_\nu \frac{d\mathcal{C}}{ds^2}\right)_{\frac{q\alpha}{2} + q'\left(-1 + \frac{\alpha}{2}\right)} \gamma_5 \lambda_\beta \, \mathrm{diag}(\bar{\sigma}_i) \, \mathrm{diag}\left(\frac{-\slashed{q}' + M(q')}{q'^2 + M^2(q')}\right) \delta(q' + q'') \times \right.
$$

$$
\left. \times \mathcal{C}\left(\frac{q'' - q'''}{2}\right) \mathrm{i}\, \gamma_5 \lambda_\gamma \, \delta(q'' + q''' - \ell) \, \mathrm{diag}\left(\frac{-\slashed{q}''' + M(q''')}{q'''^2 + M^2(q''')}\right) \delta(q''' + q) \right\}
$$

$$
= -\frac{\mathrm{i}}{2} \mathrm{tr} \left\{ \int_0^1 d\alpha \left(s_\nu \frac{d\mathcal{C}}{ds^2}\right)_{-\frac{q'-\ell}{2}\alpha + q'\left(-1 + \frac{\alpha}{2}\right)} \lambda_\beta^{ij} \bar{\sigma}_j \frac{\slashed{q}' + M_j(q')}{q'^2 + M_j^2(q')} \mathcal{C}\left(q' + \frac{\ell}{2}\right) \lambda_\gamma^{ji} \frac{-(\slashed{q}' + \slashed{\ell}) + M_i(q' + \ell)}{(q'+\ell)^2 + M_i^2(q'+\ell)} \right\}
$$

$$
= -2\mathrm{i} \lambda_\beta^{ij} \lambda_\gamma^{ji} \bar{\sigma}_j \, \mathrm{tr} \left\{ \int_0^1 d\alpha \left(s_\nu \frac{d\mathcal{C}}{ds^2}\right)_{-\frac{\ell\alpha}{2} - q'} \mathcal{C}\left(q' + \frac{\ell}{2}\right) \frac{q'(q'+\ell) + M_j(q') M_i(q'+\ell)}{\left(q'^2 + M_j^2(q')\right)\left((q'+\ell)^2 + M_i^2(q'+\ell)\right)} \right\}
$$

$$
= 2\mathrm{i} \lambda_\beta^{ij} \lambda_\gamma^{ji} \bar{\sigma}_j \, \mathrm{tr} \left\{ \int_0^1 d\alpha\, q_\nu \frac{d\mathcal{C}}{dq^2} \mathcal{C}\left(q - \frac{\ell}{2}\alpha\right) \frac{\left(q + \frac{\ell}{2}(1-\alpha)\right)\left(q - \frac{\ell}{2}(1+\alpha)\right) + M_j(q_\alpha^+) M_i(q_\alpha^-)}{\left(q_\alpha^{+2} + M_j^2(q_\alpha^+)\right)\left(q_\alpha^{-2} + M_i^2(q_\alpha^-)\right)} \right\},
$$

where q_α^+, q_α^- are defined in Eq. (3.2.29). In the same manner, one obtains for the third summand of the second term

$$
\frac{\mathrm{i}}{2} \mathrm{Tr} \left\{ d\alpha \left(s_\nu \frac{d\mathcal{C}}{ds^2}\right)_{\frac{q'\alpha}{2} - (q'+\ell)\left(-1 + \frac{\alpha}{2}\right)} \bar{\sigma}_i \lambda_\beta^{ij} \frac{\slashed{q}' + M_j(q')}{q'^2 + M_j^2(q')} \mathcal{C}\left(q' + \frac{\ell}{2}\right) \lambda_\gamma^{ji} \frac{-(\slashed{q}' + \slashed{\ell}) + M_i(q' + \ell)}{(q'+\ell)^2 + M_i^2(q'+\ell)} \right\}
$$

$$
2\mathrm{i}\, \mathrm{tr} \left\{ \int_0^1 d\alpha\, q_\nu \frac{d\mathcal{C}}{dq^2} \bar{\sigma}_i \lambda_\beta^{ij} \lambda_\gamma^{ji} \mathcal{C}\left(q - \frac{\ell\alpha}{2}\right) \frac{q_\alpha^- \cdot q_\alpha^+ + M_j(q_\alpha^-) M_i(q_\alpha^+)}{\left(q_\alpha^{-2} + M_j^2(q_\alpha^-)\right)\left(q_\alpha^{+2} + M_i^2(q_\alpha^+)\right)} \right\}.
$$

Finally, summing up all contributions, we may write

$$
\langle 0 | J_{A,\alpha}^\mu(0) | \phi_\beta(p) \rangle = 2\mathrm{i} \left(\lambda_\alpha^{ij} \lambda_\beta^{ji} + \lambda_\beta^{ij} \lambda_\alpha^{ji} \right) \mathrm{tr} \left\{ \int_0^1 d\alpha\, q_\mu \frac{d\mathcal{C}}{dq^2} \frac{M_i(q_\alpha^+)}{q_\alpha^{+2} + M_i^2(q_\alpha^+)} \right\} +
$$

$$
+ 2\mathrm{i} \left(\lambda_\alpha^{ij} \lambda_\beta^{ji} + \lambda_\beta^{ij} \lambda_\alpha^{ji} \right) \mathrm{tr} \left\{ \mathcal{C}(q) \frac{q_\mu^+ M_i(q^-)}{\left(q^{+2} + M_j^2(q^+)\right)\left(q^{-2} + M_i^2(q^-)\right)} \right\} +
$$

$$
+ 2\mathrm{i}\, \bar{\sigma}_j \left(\lambda_\alpha^{ij} \lambda_\beta^{ji} + \lambda_\beta^{ij} \lambda_\alpha^{ji} \right) \times
$$

$$
\times \mathrm{tr} \left\{ \int_0^1 d\alpha\, q_\mu \frac{d\mathcal{C}}{dq^2} \mathcal{C}\left(q - \frac{p}{2}\alpha\right) \frac{q_\alpha^+ \cdot q_\alpha^- + M_j(q_\alpha^+) M_i(q_\alpha^-)}{\left(q_\alpha^{+2} + M_j^2(q_\alpha^+)\right)\left(q_\alpha^{-2} + M_i^2(q_\alpha^-)\right)} \right\},
$$

(D.2)

with

$$q_\alpha^+ = q + \frac{p}{2}(1 - \alpha), \qquad q_\alpha^- = q - \frac{p}{2}(1 + \alpha)$$
$$q^+ = q + \frac{p}{2}, \qquad q^- = q - \frac{p}{2}.$$

(3.2.29)

Now, the decay constants can be derived from the expression (D.2) and their definitions, Eq. (3.3.49), by contraction with p^μ, hence

$$f_{\alpha\beta} = i\, p_\mu \langle 0| J_{A,\alpha}^\mu(0) |\phi_\beta(p)\rangle \frac{Z_\phi^{-1/2}}{m_\phi^2}, \qquad (D.3)$$

evaluated at the corresponding mass $p^2 = -m_\phi^2$. Owing to the properties of the Gell-Mann matrices one has $f_{\alpha\beta} = \delta_{\alpha\beta} f_\phi$ with $\phi = \pi$ for $\alpha \in \{1,2,3\}$ and $\phi = K$ for $\alpha \in \{4,5,6,7\}$. On the other hand, for the 0- and 8-component we obtain

$$f_{88}(p^2) = \frac{4}{3}\left[2f_{ss}(p^2) + f_{uu}(p^2)\right]$$
$$f_{00}(p^2) = \frac{4}{3}\left[2f_{uu}(p^2) + f_{ss}(p^2)\right]$$
$$f_{08}(p^2) = f_{80}(p^2) = \frac{4\sqrt{2}}{3}\left[f_{uu}(p^2) - f_{ss}(p^2)\right].$$

This concludes our rather lengthy calculation of the decay constants.

E Haar Measure in SU(N)

In this appendix section we are going to derive explicitly an expression for the SU(N) Haar measure appearing in the Polyakov potential in Eq. (5.2.20). In order to motivate the integration over a (Lie) group we start with the expression $\sum_{a=1}^{g} F(\hat{G}_a)$ for a discrete and finite group of order g. Such an expression is invariant with respect to replacements of the form $\hat{G}_a \to \hat{G}_c = \hat{G}_b \hat{G}_a$ with $\hat{G}_b$ fixed but arbitrary, if $\hat{G}_a$ runs over the whole group. Now, turning our attention to a continuous group G, we have to replace the summation by an integral, $\int \mu(a) f(a)\, da$, wherein $a = (a_1, \ldots, a_n)$ collects all parameters of the group, and the *measure function* $|\mu(a)|$ has to be chosen such that

$$\int_G f(a)\mu(a)\, \mathrm{d}^n a = \int_G f(c)\mu(a)\, \mathrm{d}^n a \,, \tag{E.1}$$

with c defined according to $\hat{G}(c) = \hat{G}(a)\hat{G}(b)$ and $\hat{G}(b)$ is, again, a fixed group element. In addition to Eq. (E.1) one has trivially for another parameter set $c = (c_1, \ldots, c_n)$

$$\int_G f(a)\mu(a)\, \mathrm{d}^n a = \int_G f(c)\mu(c)\, \mathrm{d}^n c \,. \tag{E.2}$$

The right-hand side of Eq. (E.2) can be transformed by standard means of functional analysis into an integral over the parameters a,

$$\int_G f(c)\mu(c)\, \mathrm{d}^n c = \int_G f(c)\mu(c)\frac{\partial c}{\partial a}\, \mathrm{d}^n a \,,$$

where the Jacobian $\mathcal{J} = \frac{\partial c}{\partial a}$ was introduced. Then, a comparison of the expressions Eqs. (E.1) and (E.2) leads to a conditional equation for the *Haar measure*:

$$\mu(a) = \frac{\partial c}{\partial a}\mu(c) \,, \tag{E.3}$$

which holds for all a. Once having calculated the Haar measure μ one can determine the volume of a (compact) group, $V(G)$ by integration over the whole parameter range,

$$V(G) = \int_G \mu(a)\, \mathrm{d}^n a \,.$$

After these introductory remarks on group integration we turn our attention to the special case of SU(N). From the definition of the Haar measure, Eq. (E.3), it is clear that one basically has to calculate the Jacobian of a special representation of the Lie group. This is generally a nontrivial matter, but we will show that in the special case of SU(N) it can be carried out in a very elegant way. Before doing this, some preparation is needed. Since a Lie group is a differentiable manifold its volume measure can be calculated from the metric tensor g: it should be well-known from differential geometry that the metric tensor corresponding to an n-dimensional submanifold $M = M(x_1, \ldots, x_n)$ is given by the squared Jacobian, $g = \mathcal{J}^\top \mathcal{J}$. Then, the n-dimensional volume measure is $\mu_V = \sqrt{\det g}$ and the total volume of the submanifold is determined according to $V(M) = \int \mu_V \, dx_1 \ldots dx_n$. Furthermore, one may introduce the following bilinear differential form

$$(ds)^2 = \text{Tr}\left[d\vec{x}^\top g \, d\vec{x} \right]. \tag{E.4}$$

It turns out that this last quantity provides a feasible method for the calculation of the Haar measure.

We now come to the Lie groups SU(N). It should be well-known that each element $M \in$ SU(N) fulfills $M^\dagger M = 1$ and $\det M = 1$. In particular, the dimension of SU(N) is $N^2 - 1$. One can say, hence, that SU(N) is a $(N^2 - 1)$-dimensional submanifold in a higher-dimensional representation space. Let us now consider an arbitrary $M \in$ SU(N) and, as a slight generalization of Eq. (E.4), the bilinear differential form (so-called Hilbert-Schmidt distance)

$$(ds)^2 = \text{Tr}\left[dM^\dagger \, dM \right] \tag{E.4$'$}$$

that is invariant under left or right multiplication of M with a group element of SU(N). If M is parametrized by $\theta_1, \ldots, \theta_{N^2-1}$, then, analogously to Eq. (E.4), the matrix elements of the metric tensor g are encoded in

$$(ds)^2 = g_{ij} \, d\theta^i \, d\theta^j. \tag{E.5}$$

Hence, the goal in the following is the calculation of ds which allows for the determination of the metric tensor g and consequently the evaluation of the volume measure $\sqrt{\det g}$.

It is well-known from linear algebra that each element of a unitary group, in particular an element of SU(N), can be diagonalized using a unitary matrix $U \in$ SU(N):

$$M = U \Lambda U^\dagger,$$

where $\Lambda = \text{diag}(\Lambda_1, \ldots, \Lambda_N)$ is a $N \times N$ diagonal matrix with $|\Lambda_i| = 1$ for $i \in \{1, \ldots, N\}$, and $U^\dagger = U^{-1}$. Having this, one obtains

$$dM = U\left(d\Lambda + U^{-1} \, dU \, \Lambda - \Lambda U^{-1} \, dU \right) U^{-1}$$
$$dM^\dagger = U\left(d\Lambda^\dagger + U^{-1} \, dU \, \Lambda^\dagger - \Lambda^\dagger U^{-1} \, dU \right) U^{-1}.$$

Inserting these into the bilinear form Eq. (E.4$'$) leads, after a straightforward calculation, to

$$(ds)^2 = \sum_{i=1}^N |d\Lambda_i|^2 - 2\sum_{i<\ell} |\Lambda_\ell - \Lambda_i|^2 \left(U^{-1} \, dU \right)_{i\ell} \left(U^{-1} \, dU \right)_{\ell i}.$$

From $U^\dagger U = 1$ one derives $(U^{-1} \, dU)_{\ell i} = -(U^{-1} \, dU)^*_{i\ell}$, and therefore we finally may write

$$(ds)^2 = \sum_{i=1}^N |d\Lambda_i|^2 + 2\sum_{i<\ell} |\Lambda_\ell - \Lambda_i|^2 \left| \left(U^{-1} \, dU \right)_{i\ell} \right|^2. \tag{E.6}$$

In the case of SU(N) not all of the Λ_i are independent since the determinant is subject to the constraint $\det M = \prod_{i=1}^{N} \Lambda_i = 1$. This means, that Λ_N, say, may be written as $\Lambda_N = \prod_{i=1}^{N-1} \Lambda_i^{-1}$ and, consequently,

$$\mathrm{d}\Lambda_N = -\left(\prod_{i=1}^{N-1} \Lambda_i^{-1}\right) \sum_{j=1}^{N-1} \frac{\mathrm{d}\Lambda_j}{\Lambda_j}.$$

This allows to eliminate $\mathrm{d}\Lambda_N$ in Eq. (E.6) resulting in

$$\sum_{i=1}^{N} |\mathrm{d}\Lambda_i|^2 = \sum_{i=1}^{N-1} 2|\mathrm{d}\Lambda_i|^2 + \sum_{i\neq j}^{N-1} \mathrm{d}\Lambda_i \, \mathrm{d}\Lambda_j, \qquad (E.7)$$

because $|\Lambda_i| = 1$.

At this point we are almost done, since the basic results can now be read off from Eqs. (E.6) and (E.7): First, Eq. (E.6) supposes to introduce the diagonal elements Λ_i and the elements $\left(U^{-1}\,\mathrm{d}U\right)_{i<\ell}$ as new group parameters. Note, in particular, that from the left translational invariance of the Lie algebra su(N) it follows immediately that $U^{-1}\,\mathrm{d}U$ is an element of the Lie algebra su(N), too. Since the parameters are assumed to be real, one should use real and imaginary parts, $\mathrm{Re}\left(U^{-1}\,\mathrm{d}U\right)_{i<\ell}$ and $\mathrm{Im}\left(U^{-1}\,\mathrm{d}U\right)_{i<\ell}$, respectively. In addition, taking into account that the Λ_i represent $N-1$ independent parameters, one has totally $(N-1)+2\cdot\frac{1}{2}(N-1)N = N^2 - 1$ parameters, as it has to be. Second, Eq. (E.6) shows explicitly that $\mathrm{d}\Lambda_i$ and $\left(U^{-1}\,\mathrm{d}U\right)_{i<\ell}$ decouple in the sense that there are no terms of the form $\mathrm{d}\Lambda_i\left(U^{-1}\,\mathrm{d}U\right)_{j\ell}$. This means, that the SU(N) Haar measure may be decomposed into a *product measure*,

$$\mu_{\mathrm{SU}(N)}\left(\Lambda_i; \left(U^{-1}\,\mathrm{d}U\right)_{i<\ell}\right) = \mu_\Lambda(\Lambda_1,\ldots,\Lambda_{N-1}) \otimes \mu_U\left(\left(U^{-1}\,\mathrm{d}U\right)_{i<\ell}\right).$$

As a consequence, the determinants necessary for a calculation of the metric tensor, g, factorize, $\det g = \det g_\Lambda \cdot \det g_U$. The form of the matrix g_Λ can easily be read off from Eqs. (E.6) and (E.7),

$$g_\Lambda = \begin{pmatrix} 2 & 1 & & \cdots & & 1 \\ 1 & 2 & 1 & \cdots & & 1 \\ & & \cdots\cdots & & & \\ 1 & & \cdots & & 1 & 2 \end{pmatrix}$$

and its determinant is readily evaluated to be $\det g_\Lambda = N$. Thus, the (differential) SU(N) Haar measure whose weighting factor is $\sqrt{\det g} = \sqrt{\det g_\Lambda} \cdot \sqrt{\det g_U}$ reads

$$\mathrm{d}\mu_{\mathrm{SU}(N)} = \sqrt{N} \prod_{i=1}^{N-1} |\mathrm{d}\Lambda_i| \prod_{1\leq i<\ell\leq N} |\Lambda_i - \Lambda_\ell|^2 \prod_{1\leq i<\ell\leq N} 2\,\mathrm{Re}\left(U^{-1}\,\mathrm{d}U\right)_{i\ell} \mathrm{Im}\left(U^{-1}\,\mathrm{d}U\right)_{i\ell} \qquad (E.8)$$

Since the last term does not depend on the Λ_i, it can be integrated out to give an unimportant factor, thence, in what follows it will simply be discarded. Therefore, the *weighting factor* of the Haar measure is determined by the *Vandermonde determinant* of a diagonalized SU(N) matrix.

According to the definition (E.3) we write

$$\mu(\Lambda_1, \ldots, \Lambda_{N-1})_{\mathrm{SU}(N)} = \prod_{1 \le i < \ell \le N} |\Lambda_i - \Lambda_\ell|^2 \, , \tag{E.9}$$

where, as already mentioned above, $\Lambda_N = \prod_{i=1}^{N-1} \Lambda_i^{-1}$.

Example: Haar Measure of SU(3)

As an application of the result derived so far we evaluate the Haar measure (E.9) for SU(3). Since any SU(3) matrix has determinant 1, a diagonal matrix may be parametrized as follows:

$$\Lambda = \mathrm{diag}\left(e^{-i\theta_1}, e^{-i\theta_2}, e^{i(\theta_1 + \theta_2)}\right) \, .$$

The Vandermonde determinant is readily evaluated to

$$\prod_{1 \le i < \ell \le 3} |\Lambda_i - \Lambda_\ell|^2 = 4\Big(\sin(\theta_1 - \theta_2) + \sin(\theta_1 + 2\theta_2) - \sin(2\theta_1 + \theta_2) \Big)^2 \, .$$

If we set $\theta_1 - \theta_2 = 2\phi_3$ and $2\theta_1 + \theta_2 = \phi_3 + \sqrt{3}\phi_8$, the above expression can be written as

$$\mu_\Lambda(\phi_3, \phi_8) = \prod_{1 \le i < \ell \le 3} |\Lambda_i - \Lambda_\ell|^2 = 4\left(\sin(2\phi_3) - \sin(\phi_3 + \sqrt{3}\phi_8) + \sin(-\phi_3 + \sqrt{3}\phi_8) \right)^2 \, .$$

It turns out, that in order to cover the whole SU(3) manifold the integration ranges have to be chosen as $-\pi \le \frac{\phi_3}{2} \le \pi$ and $-\pi \le \frac{1}{\sqrt{3}}\phi_8 \le \pi$. Then it easy to determine the volume of SU(3),

$$\mu(\mathrm{SU}(3)) = 24\pi^2 \, , \tag{E.10}$$

and the normalized Haar measure is

$$\frac{\mu(\phi_3, \phi_8)}{\mu(\mathrm{SU}(3))} = \frac{1}{6\pi^2} \left(\sin(2\phi_3) - \sin(\phi_3 + \sqrt{3}\phi_8) + \sin(-\phi_3 + \sqrt{3}\phi_8) \right)^2 \, , \tag{E.11}$$

which is the result used for the Polyakov potential.

Bibliography

[1] V. Baluni, "CP-violating effects in QCD," *Phys. Rev.*, vol. D19, pp. 2227–2230, 1979.

[2] R. J. Crewther, P. Di Vecchia, G. Veneziano, and E. Witten, "Chiral estimate of the electric dipole moment of the neutron in quantum chromodynamics," *Phys. Lett.*, vol. B88, p. 123, 1979.

[3] C. A. Baker *et al.*, "An improved experimental limit on the electric dipole moment of the neutron," *Phys. Rev. Lett.*, vol. 97, p. 131801, 2006.

[4] R. D. Peccei and H. R. Quinn, "CP-conservation in the presence of pseudoparticles," *Phys. Rev. Lett.*, vol. 38, pp. 1440–1443, Jun 1977.

[5] R. D. Peccei and H. R. Quinn, "Constraints imposed by CP-conservation in the presence of pseudoparticles," *Phys. Rev. D*, vol. 16, pp. 1791–1797, Sep 1977.

[6] F. Wilczek, "Problem of strong P and T invariance in the presence of instantons," *Phys. Rev. Lett.*, vol. 40, pp. 279–282, Jan 1978.

[7] S. Weinberg, "A new light boson?," *Phys. Rev. Lett.*, vol. 40, pp. 223–226, Jan 1978.

[8] R. P. Feynman and A. R. Hibbs, *Quantum Mechanics and Path Integrals*. International Series in Pure and Applied Physics, New York, U.S.A.: McGraw-Hill, 1965.

[9] L. H. Ryder, *Quantum Field Theory – 2nd ed.* Cambridge, UK: Cambridge University Press, 1996.

[10] C. Itzykson and J.-B. Zuber, *Quantum Field Theory.* New York: McGraw-Hill Book Co, 1985.

[11] V. N. Gribov, "Quantization of non-Abelian gauge theories," *Nucl. Phys.*, vol. B139, p. 1, 1978.

[12] L. D. Faddeev and V. N. Popov, "Feynman diagrams for the Yang-Mills field," *Phys. Lett.*, vol. B25, pp. 29–30, 1967.

[13] G. 't Hooft and M. J. G. Veltman, "Regularization and renormalization of gauge fields," *Nucl. Phys.*, vol. B44, pp. 189–213, 1972.

[14] D. J. Gross and F. Wilczek, "Ultraviolet behavior of non-Abelian gauge theories," *Phys. Rev. Lett.*, vol. 30, pp. 1343–1346, 1973.

[15] D. J. Gross and F. Wilczek, "Asymptotically free gauge theories. 1," *Phys. Rev.*, vol. D8, pp. 3633–3652, 1973.

[16] H. D. Politzer, "Reliable perturbative results for strong interactions?," *Phys. Rev. Lett.*, vol. 30, pp. 1346–1349, 1973.

[17] H. D. Politzer, "Asymptotic freedom: an approach to strong interactions," *Phys. Rept.*, vol. 14, pp. 129–180, 1974.

[18] A. W. Thomas and W. Weise, *The Structure of the Nucleon*. Berlin, Germany: Wiley-VCH Verlag, 2001.

[19] C. Amsler *et al.*, "Review of particle physics," *Phys. Lett.*, vol. B667, p. 1, 2008.

[20] M. Gell-Mann, R. J. Oakes, and B. Renner, "Behavior of current divergences under $SU(3) \times SU(3)$," *Phys. Rev.*, vol. 175, pp. 2195–2199, Nov 1968.

[21] M. L. Goldberger and S. B. Treiman, "Decay of the pi-meson," *Phys. Rev.*, vol. 110, pp. 1178–1184, Jun 1958.

[22] K. G. Wilson, "Confinement of quarks," *Phys. Rev. D*, vol. 10, pp. 2445–2459, Oct 1974.

[23] J. B. Kogut and L. Susskind, "Hamiltonian formulation of Wilson's lattice gauge theories," *Phys. Rev.*, vol. D11, p. 395, 1975.

[24] H. J. Rothe, *Lattice Gauge Theories—An Introduction*. Singapore: World Scientific Lecture Notes in Physics – Vol. 74, 2005.

[25] M. Creutz, "The evil that is rooting," *Phys. Lett.*, vol. B649, pp. 230–234, 2007.

[26] H. B. Nielsen and M. Ninomiya, "Absence of neutrinos on a lattice. 1. Proof by homotopy theory," *Nucl. Phys.*, vol. B185, p. 20, 1981.

[27] C. D. Roberts and S. M. Schmidt, "Dyson-Schwinger equations: density, temperature and continuum strong QCD," *Prog. Part. Nucl. Phys.*, vol. 45, pp. S1–S103, 2000.

[28] J. Praschifka, C. D. Roberts, and R. T. Cahill, "QCD bosonization and the meson effective action," *Phys. Rev. D*, vol. 36, pp. 209–220, Jul 1987.

[29] Y. Nambu and G. Jona-Lasinio, "Dynamical model of elementary particles based on an analogy with superconductivity. I," *Phys. Rev.*, vol. 122, pp. 345–358, 1961.

[30] Y. Nambu and G. Jona-Lasinio, "Dynamical model of elementary particles based on an analogy with superconductivity. II," *Phys. Rev.*, vol. 124, pp. 246–254, 1961.

[31] V. Bernard, R. L. Jaffe, and U. G. Meissner, "Strangeness mixing and quenching in the Nambu–Jona-Lasinio model," *Nucl. Phys.*, vol. B308, p. 753, 1988.

[32] M. Kobayashi and T. Maskawa, "CP-violation in the renormalizable theory of weak interaction," *Prog. Theor. Phys.*, vol. 49, pp. 652–657, 1973.

[33] G. 't Hooft, "Symmetry breaking through Bell-Jackiw anomalies," *Phys. Rev. Lett.*, vol. 37, p. 8, Jul 1976.

[34] U. Vogl and W. Weise, "The Nambu and Jona-Lasinio model: its implications for hadrons and nuclei," *Prog. Part. Nucl. Phys.*, vol. 27, pp. 195–272, 1991.

[35] C. Ratti, M. A. Thaler, and W. Weise, "Phases of QCD: lattice thermodynamics and a field theoretical model," *Phys. Rev.*, vol. D73, p. 014019, 2006.

[36] S. Rößner, "Field theoretical modeling of the QCD phase diagram," *Diploma Thesis*, 2006.

[37] S. Rößner, C. Ratti, and W. Weise, "Polyakov loop, diquarks and the two-flavor phase diagram," *Phys. Rev.*, vol. D75, p. 034007, 2007.

[38] S. Rößner, T. Hell, C. Ratti, and W. Weise, "The chiral and deconfinement crossover transitions: PNJL model beyond mean field," *Nucl. Phys.*, vol. A814, pp. 118–143, 2008.

[39] T. Hell, S. Rößner, M. Cristoforetti, and W. Weise, "Dynamics and thermodynamics of a nonlocal PNJL model with running coupling," *Phys. Rev.*, vol. D79, p. 014022, 2009.

[40] T. Hell, S. Rößner, M. Cristoforetti, and W. Weise, "Thermodynamics of a three-flavor nonlocal Polyakov–Nambu–Jona-Lasinio model," *Phys. Rev.*, vol. D81, p. 074034, 2010.

[41] C. D. Roberts and A. G. Williams, "Dyson-Schwinger equations and their application to hadronic physics," *Prog. Part. Nucl. Phys.*, vol. 33, pp. 477–575, 1994.

[42] C. D. Roberts, "Hadron properties and Dyson-Schwinger equations," *Prog. Part. Nucl. Phys.*, vol. 61, pp. 50–65, 2008.

[43] R. Alkofer and L. v. Smekal, "The infrared behavior of QCD Green's functions: confinement, dynamical symmetry breaking, and hadrons as relativistic bound states," *Phys. Rept.*, vol. 353, p. 281, 2001.

[44] C. S. Fischer, "Infrared properties of QCD from Dyson-Schwinger equations," *J. Phys.*, vol. G32, pp. R253–R291, 2006.

[45] C. J. Burden, L. Qian, C. D. Roberts, P. C. Tandy, and M. J. Thomson, "Ground-state spectrum of light-quark mesons," *Phys. Rev.*, vol. C55, pp. 2649–2664, 1997.

[46] R. D. Bowler and M. C. Birse, "A nonlocal, covariant generalization of the NJL model," *Nucl. Phys.*, vol. A582, pp. 655–664, 1995.

[47] R. S. Plant and M. C. Birse, "Meson properties in an extended nonlocal NJL model," *Nucl. Phys.*, vol. A628, pp. 607–644, 1998.

[48] D. Gomez Dumm, A. G. Grunfeld, and N. N. Scoccola, "On covariant nonlocal chiral quark models with separable interactions," *Phys. Rev.*, vol. D74, p. 054026, 2006.

[49] J. W. Bos, J. H. Koch, and H. W. L. Naus, "Currents and Ward-Takahashi identities for nonlocal quantum field theories," *Phys. Rev.*, vol. C44, pp. 485–490, 1991.

[50] H. D. Politzer, "Effective quark masses in the chiral limit," *Nucl. Phys.*, vol. B117, p. 397, 1976.

[51] V. A. Miransky, "On dynamical chiral symmetry breaking," *Phys. Lett.*, vol. B165, pp. 401–404, 1985.

[52] M. Lavelle and D. McMullan, "Constituent quarks from QCD," *Phys. Rept.*, vol. 279, pp. 1–65, 1997.

[53] P. O. Bowman, U. M. Heller, D. B. Leinweber, and A. G. Williams, "Modeling the quark propagator," 2002.

[54] S. P. Klevansky, "The Nambu–Jona-Lasinio model of quantum chromodynamics," *Rev. Mod. Phys.*, vol. 64, pp. 649–708, 1992.

[55] S. L. Adler, "Axial vector vertex in spinor electrodynamics," *Phys. Rev.*, vol. 177, pp. 2426–2438, 1969.

[56] J. S. Bell and R. Jackiw, "A PCAC puzzle: $\pi_0 \to \gamma\gamma$ in the sigma model," *Nuovo Cim.*, vol. A60, pp. 47–61, 1969.

[57] K. Fujikawa, "Path-integral measure for gauge invariant fermion theories," *Phys. Rev. Lett.*, vol. 42, pp. 1195–1198, Apr 1979.

[58] T. Schäfer and E. V. Shuryak, "Instantons in QCD," *Rev. Mod. Phys.*, vol. 70, pp. 323–426, 1998.

[59] A. A. Belavin, A. M. Polyakov, A. S. Schwartz, and Y. S. Tyupkin, "Pseudoparticle solutions of the Yang-Mills equations," *Phys. Lett.*, vol. B59, pp. 85–87, 1975.

[60] T. Banks and A. Casher, "Chiral symmetry breaking in confining theories," *Nucl. Phys.*, vol. B169, p. 103, 1980.

[61] R. E. Cutkosky, P. V. Landshoff, D. I. Olive, and J. C. Polkinghorne, "A nonanalytic S-matrix," *Nucl. Phys.*, vol. B12, pp. 281–300, 1969.

[62] A. Scarpettini, D. Gomez Dumm, and N. N. Scoccola, "Light pseudoscalar mesons in a nonlocal SU(3) chiral quark model," *Phys. Rev.*, vol. D69, p. 114018, 2004.

[63] B. Di Micco, "Eta–eta-prime mixing angle and eta-prime gluonium content extraction from the KLOE R(phi) measurement," *Eur. Phys. J.*, vol. A38, pp. 129–131, 2008.

[64] S. Klimt, M. Lutz, U. Vogl, and W. Weise, "Generalized SU(3) Nambu–Jona-Lasinio model. Part 1. Mesonic modes," *Nucl. Phys.*, vol. A516, pp. 429–468, 1990.

[65] S. Klimt, M. Lutz, and W. Weise, "Chiral phase transition in the SU(3) Nambu and Jona-Lasinio model," *Phys. Lett.*, vol. B249, pp. 386–390, 1990.

[66] P. Rehberg, S. P. Klevansky, and J. Hüfner, "Hadronization in the SU(3) Nambu–Jona-Lasinio model," *Phys. Rev.*, vol. C53, pp. 410–429, 1996.

[67] C. Ratti, S. Rößner, and W. Weise, "Quark-number susceptibilities: lattice QCD versus PNJL model," *Phys. Lett.*, vol. B649, pp. 57–60, 2007.

[68] F. R. Brown *et al.*, "On the existence of a phase transition for QCD with three light quarks," *Phys. Rev. Lett.*, vol. 65, pp. 2491–2494, 1990.

[69] E. Laermann and O. Philipsen, "Status of lattice QCD at finite temperature," *Ann. Rev. Nucl. Part. Sci.*, vol. 53, pp. 163–198, 2003.

[70] R. D. Pisarski and F. Wilczek, "Remarks on the chiral phase transition in chromodynamics," *Phys. Rev.*, vol. D29, pp. 338–341, 1984.

[71] J. B. Natowitz *et al.*, "Limiting temperatures and the equation of state of nuclear matter," *Phys. Rev. Lett.*, vol. 89, p. 212701, 2002.

[72] V. A. Karnaukhov *et al.*, "Critical temperature for the nuclear liquid-gas phase transition (from multifragmentation and fission)," *Phys. Atom. Nucl.*, vol. 71, pp. 2067–2073, 2008.

[73] V. A. Karnaukhov *et al.*, "Multifragmentation and nuclear phase transitions (liquid-fog and liquid-gas)," *Nucl. Phys.*, vol. A734, pp. 520–523, 2004.

[74] M. G. Alford, "Color superconductivity in ultra-dense quark matter," *PoS*, vol. LAT2006, p. 001, 2006.

[75] M. Cheng *et al.*, "The QCD equation of state with almost physical quark masses," *Phys. Rev.*, vol. D77, p. 014511, 2008.

[76] M. Cheng *et al.*, "The finite temperature QCD using $2 + 1$ flavors of domain-wall fermions at $N_t = 8$," *Phys. Rev.*, vol. D81, p. 054510, 2010.

[77] Y. Aoki *et al.*, "The QCD transition temperature: results with physical masses in the continuum limit II," *JHEP*, vol. 06, p. 088, 2009.

[78] L. P. Csernai, J. I. Kapusta, and L. D. McLerran, "On the strongly interacting low-viscosity matter created in relativistic nuclear collisions," *Phys. Rev. Lett.*, vol. 97, p. 152303, 2006.

[79] P. Kovtun, D. T. Son, and A. O. Starinets, "Viscosity in strongly interacting quantum field theories from black hole physics," *Phys. Rev. Lett.*, vol. 94, p. 111601, 2005.

[80] L. McLerran and R. D. Pisarski, "Phases of cold, dense quarks at large N_c," *Nucl. Phys.*, vol. A796, pp. 83–100, 2007.

[81] Y. Hidaka, L. D. McLerran, and R. D. Pisarski, "Baryons and the phase diagram for a large number of colors and flavors," *Nucl. Phys.*, vol. A808, pp. 117–123, 2008.

[82] A. Andronic, P. Braun-Munzinger, and J. Stachel, "Thermal hadron production in relativistic nuclear collisions: the sigma meson, the horn, and the QCD phase transition," *Phys. Lett.*, vol. B673, p. 142, 2009.

[83] A. Andronic, P. Braun-Munzinger, and J. Stachel, "Thermal hadron production in relativistic nuclear collisions," *Acta Phys. Polon.*, vol. B40, pp. 1005–1012, 2009.

[84] M. Cristoforetti, T. Hell, B. Klein, and W. Weise, "Thermodynamics and quark susceptibilities: a Monte Carlo approach to the PNJL model," 2010.

[85] M. Stephanov, K. Rajagopal, and E. Shuryak, "Event-by-event fluctuations in heavy-ion collisions and the qcd critical point," *Phys. Rev. D*, vol. 60, p. 114028, Nov 1999.

[86] B. Berdnikov and K. Rajagopal, "Slowing out of equilibrium near the QCD critical point," *Phys. Rev. D*, vol. 61, no. 10, p. 105017, 2000.

[87] D. T. Son and M. A. Stephanov, "Dynamic universality class of the QCD critical point," *Phys. Rev.*, vol. D70, p. 056001, 2004.

[88] P. C. Hohenberg and B. I. Halperin, "Theory of dynamic critical phenomena," *Rev. Mod. Phys.*, vol. 49, no. 3, pp. 435–479, 1977.

[89] Z. Fodor and S. D. Katz, "Lattice determination of the critical point of QCD at finite T and μ," *JHEP*, vol. 03, p. 014, 2002.

[90] Z. Fodor and S. D. Katz, "Critical point of QCD at finite T and μ, lattice results for physical quark masses," *JHEP*, vol. 04, p. 050, 2004.

[91] P. de Forcrand and O. Philipsen, "The QCD phase diagram for three degenerate flavors and small baryon density," *Nucl. Phys.*, vol. B673, pp. 170–186, 2003.

[92] P. de Forcrand and O. Philipsen, "The phase diagram of $N_f = 3$ QCD for small baryon densities," *Nucl. Phys. Proc. Suppl.*, vol. 129, pp. 521–523, 2004.

[93] P. de Forcrand and O. Philipsen, "The chiral critical line of $N_f = 2 + 1$ QCD at zero and nonzero baryon density," *JHEP*, vol. 01, p. 077, 2007.

[94] S. Chandrasekharan and A. C. Mehta, "Effects of the anomaly on the two-flavor QCD chiral phase transition," *Phys. Rev. Lett.*, vol. 99, p. 142004, 2007.

[95] Y. Aoki, Z. Fodor, S. D. Katz, and K. K. Szabo, "The QCD transition temperature: results with physical masses in the continuum limit," *Phys. Lett.*, vol. B643, pp. 46–54, 2006.

[96] A. Bazavov *et al.*, "Equation of state and QCD transition at finite temperature," *Phys. Rev.*, vol. D80, p. 014504, 2009.

[97] L. D. McLerran and B. Svetitsky, "Quark liberation at high temperature: a Monte Carlo study of SU(2) gauge theory," *Phys. Rev.*, vol. D24, p. 450, 1981.

[98] N. Weiss, "The Wilson line in finite-temperature gauge theories," *Phys. Rev.*, vol. D25, p. 2667, 1982.

[99] S. Borsanyi *et al.*, "Is there still any T_c mystery in lattice QCD? Results with physical masses in the continuum limit III," 2010.

[100] M. Cheng *et al.*, "Equation of state for physical quark masses," *Phys. Rev.*, vol. D81, p. 054504, 2010.

[101] A. Bazavov and P. Petreczky, "Deconfinement and chiral transition with the highly improved staggered-quark (HISQ) action," 2010.

[102] P. Huovinen and P. Petreczky, "QCD equation of state and hadron resonance gas," *Nucl. Phys.*, vol. A837, pp. 26–53, 2010.

[103] M. A. Stephanov, "QCD phase diagram: an overview," *PoS*, vol. LAT2006, p. 024, 2006.

[104] M. Asakawa and K. Yazaki, "Chiral restoration at finite density and temperature," *Nucl. Phys.*, vol. A504, pp. 668–684, 1989.

[105] A. Barducci, R. Casalbuoni, S. De Curtis, R. Gatto, and G. Pettini, "Chiral symmetry breaking in QCD at finite temperature and density," *Phys. Lett.*, vol. B231, p. 463, 1989.

[106] A. Barducci, R. Casalbuoni, S. De Curtis, R. Gatto, and G. Pettini, "Chiral phase transition in QCD for finite temperature and density," *Phys. Rev.*, vol. D41, p. 1610, 1990.

[107] A. Barducci, R. Casalbuoni, G. Pettini, and R. Gatto, "Chiral phases of QCD at finite density and temperature," *Phys. Rev.*, vol. D49, pp. 426–436, 1994.

[108] J. Berges and K. Rajagopal, "Color superconductivity and chiral symmetry restoration at nonzero baryon density and temperature," *Nucl. Phys.*, vol. B538, pp. 215–232, 1999.

[109] A. M. Halasz, A. D. Jackson, R. E. Shrock, M. A. Stephanov, and J. J. M. Verbaarschot, "On the phase diagram of QCD," *Phys. Rev.*, vol. D58, p. 096007, 1998.

[110] O. Scavenius, A. Mocsy, I. N. Mishustin, and D. H. Rischke, "Chiral phase transition within effective models with constituent quarks," *Phys. Rev.*, vol. C64, p. 045202, 2001.

[111] N. G. Antoniou and A. S. Kapoyannis, "Bootstraping the QCD critical point," *Phys. Lett.*, vol. B563, pp. 165–172, 2003.

[112] Y. Hatta and T. Ikeda, "Universality, the QCD critical / tricritical point and the quark-number susceptibility," *Phys. Rev.*, vol. D67, p. 014028, 2003.

[113] S. Ejiri *et al.*, "Study of QCD thermodynamics at finite density by Taylor expansion," *Prog. Theor. Phys. Suppl.*, vol. 153, pp. 118–126, 2004.

[114] R. V. Gavai and S. Gupta, "The critical end point of QCD," *Phys. Rev.*, vol. D71, p. 114014, 2005.

[115] J. I. Kapusta and C. Gale, *Finite-Temperature Field Theory*. Cambridge, UK: Cambridge University Press, 2006.

[116] M. Le Bellac, *Thermal Field Theory*. Cambridge, UK: Cambridge University Press, 2000.

[117] A. M. Polyakov, "Thermal properties of gauge fields and quark liberation," 1977. IC-77-135.

[118] L. D. McLerran and B. Svetitsky, "A Monte Carlo study of SU(2) Yang-Mills theory at finite temperature," *Phys. Lett.*, vol. B98, p. 195, 1981.

[119] G. 't Hooft, "On the phase transition towards permanent quark confinement," *Nucl. Phys.*, vol. B138, p. 1, 1978.

[120] L. Susskind, "Lattice models of quark confinement at high temperature," *Phys. Rev.*, vol. D20, pp. 2610–2618, 1979.

[121] K. Fukushima, "Chiral effective model with the Polyakov loop," *Phys. Lett. B*, vol. 591, p. 277, 2004.

[122] T. Hell, "Thermodynamics of an NJL-type model with a running coupling," *Diploma Thesis*, 2007.

[123] K. Fukushima, "Relation between the Polyakov loop and the chiral order parameter at strong coupling," *Phys. Rev.*, vol. D68, p. 045004, 2003.

[124] Y. Hatta and K. Fukushima, "Linking the chiral and deconfinement phase transitions," *Phys. Rev.*, vol. D69, p. 097502, 2004.

[125] F. Karsch, E. Laermann, and A. Peikert, "Quark mass and flavor dependence of the QCD phase transition," *Nucl. Phys.*, vol. B605, pp. 579–599, 2001.

[126] B.-J. Schaefer, J. M. Pawlowski, and J. Wambach, "The phase structure of the Polyakov–quark-meson model," *Phys. Rev.*, vol. D76, p. 074023, 2007.

[127] K. Fukushima, "Phase diagrams in the three-flavor Nambu–Jona-Lasinio model with the Polyakov loop," *Phys. Rev.*, vol. D77, p. 114028, 2008.

[128] G. Boyd *et al.*, "Thermodynamics of SU(3) lattice gauge theory," *Nucl. Phys.*, vol. B469, pp. 419–444, 1996.

[129] O. Kaczmarek, F. Karsch, P. Petreczky, and F. Zantow, "Heavy quark-antiquark free energy and the renormalized Polyakov loop," *Phys. Lett.*, vol. B543, pp. 41–47, 2002.

[130] G. 't Hooft *et al.*, "Recent developments in gauge theories. Proceedings, Nato Advanced Study Institute, Cargese, France, August 26–September 8, 1979," 1980. New York, USA: Plenum (1980) 438 P. (Nato Advanced Study Institutes Series: Series B, Physics, 59).

[131] A. Casher, "Chiral symmetry breaking in quark-confining theories," *Phys. Lett.*, vol. B83, p. 395, 1979.

[132] L. Y. Glozman, "Confined but chirally symmetric hadrons at large density and the Casher's argument," *Phys. Rev.*, vol. D80, p. 037701, 2009.

[133] M. Cheng *et al.*, "The transition temperature in QCD," *Phys. Rev.*, vol. D74, p. 054507, 2006.

[134] H. Hansen *et al.*, "Mesonic correlation functions at finite temperature and density in the Nambu–Jona-Lasinio model with a Polyakov loop," *Phys. Rev.*, vol. D75, p. 065004, 2007.

[135] J. Hüfner, S. P. Klevansky, P. Zhuang, and H. Voss, "Thermodynamics of a quark plasma beyond the mean field: a generalized Beth-Uhlenbeck approach," *Annals Phys.*, vol. 234, pp. 225–244, 1994.

[136] M. Buballa, "NJL model analysis of quark matter at large density," *Phys. Rept.*, vol. 407, pp. 205–376, 2005.

[137] P. Gerber and H. Leutwyler, "Hadrons below the chiral phase transition," *Nucl. Phys.*, vol. B321, p. 387, 1989.

[138] D. Toublan, "Pion dynamics at finite temperature," *Phys. Rev.*, vol. D56, pp. 5629–5645, 1997.

[139] N. Kaiser, "Pi–pi-scattering lengths at finite temperature," *Phys. Rev.*, vol. C59, pp. 2945–2947, 1999.

[140] L. McLerran, K. Redlich, and C. Sasaki, "Quarkyonic matter and chiral symmetry breaking," *Nucl. Phys.*, vol. A824, pp. 86–100, 2009.

[141] C. R. Allton *et al.*, "Thermodynamics of two-flavor QCD to sixth order in quark chemical potential," *Phys. Rev.*, vol. D71, p. 054508, 2005.

[142] N. Kaiser, P. de Homont, and W. Weise, "In-medium chiral condensate beyond linear density approximation," *Phys. Rev.*, vol. C77, p. 025204, 2008.

[143] T. D. Cohen, R. J. Furnstahl, and D. K. Griegel, "Quark and gluon condensates in nuclear matter," *Phys. Rev.*, vol. C45, pp. 1881–1893, 1992.

[144] E. G. Drukarev and E. M. Levin, "The QCD sum rules and nuclear matter. 2," *Nucl. Phys.*, vol. A511, pp. 679–700, 1990.

[145] K.-I. Kondo, "Toward a first-principle derivation of confinement and chiral-symmetry-breaking crossover transitions in QCD," 2010.

[146] K.-I. Kondo, T. Murakami, and T. Shinohara, "Yang-Mills theory constructed from Cho-Faddeev-Niemi decomposition," *Prog. Theor. Phys.*, vol. 115, pp. 201–216, 2006.

[147] K.-I. Kondo, "Magnetic condensation, Abelian dominance and instability of Savvidy vacuum," *Phys. Lett.*, vol. B600, pp. 287–296, 2004.

[148] K.-I. Kondo, "Gauge-invariant gluon mass, infrared Abelian dominance and stability of magnetic vacuum," *Phys. Rev.*, vol. D74, p. 125003, 2006.

[149] K.-I. Kondo, T. Shinohara, and T. Murakami, "Reformulating $SU(N)$ Yang-Mills theory based on change of variables," *Prog. Theor. Phys.*, vol. 120, pp. 1–50, 2008.

[150] L. D. Faddeev and A. J. Niemi, "Partial duality in $SU(N)$ Yang-Mills theory," *Phys. Lett.*, vol. B449, pp. 214–218, 1999.

[151] Y. M. Cho, "Abelian dominance in Wilson loops," *Phys. Rev.*, vol. D62, p. 074009, 2000.

[152] W. S. Bae, Y. M. Cho, and S.-W. Kim, "QCD versus Skyrme-Faddeev theory," *Phys. Rev.*, vol. D65, p. 025005, 2002.

[153] Y. M. Cho, "A restricted gauge theory," *Phys. Rev.*, vol. D21, p. 1080, 1980.

[154] Y. M. Cho, "Glueball spectrum in extended QCD," *Phys. Rev. Lett.*, vol. 46, pp. 302–306, 1981.

[155] L. D. Faddeev and A. J. Niemi, "Partially dual variables in $SU(2)$ Yang-Mills theory," *Phys. Rev. Lett.*, vol. 82, pp. 1624–1627, 1999.

[156] L. D. Faddeev and A. J. Niemi, "Spin-charge separation, conformal covariance and the $SU(2)$ Yang-Mills theory," *Nucl. Phys.*, vol. B776, pp. 38–65, 2007.

[157] S. V. Shabanov, "An effective action for monopoles and knot solitons in Yang-Mills theory," *Phys. Lett.*, vol. B458, pp. 322–330, 1999.

[158] S. V. Shabanov, "Yang-Mills theory as an Abelian theory without gauge fixing," *Phys. Lett.*, vol. B463, pp. 263–272, 1999.

[159] Y. M. Cho, "Gauge theory on homogeneous fiber bundle," 1982. CERN-TH-3414.

[160] G. 't Hooft, "Magnetic monopoles in unified gauge theories," *Nucl. Phys.*, vol. B79, pp. 276–284, 1974.

[161] A. M. Polyakov, "Particle spectrum in quantum field theory," *JETP Lett.*, vol. 20, pp. 194–195, 1974.

[162] J. Arafune, P. G. O. Freund, and C. J. Goebel, "Topology of Higgs fields," *J. Math. Phys.*, vol. 16, p. 433, 1975.

[163] M. I. Monastyrsky and A. M. Perelomov, "Some remarks on monopoles in gauge gield theories," *JETP Lett.*, vol. 21, p. 43, 1975.

[164] A. Shibata *et al.*, "Compact lattice formulation of Cho-Faddeev-Niemi decomposition: gluon mass generation and infrared Abelian dominance," *Phys. Lett.*, vol. B653, pp. 101–108, 2007.

[165] S. Mandelstam, "Vortices and quark confinement in non-Abelian gauge theories," *Phys. Rept.*, vol. 23, pp. 245–249, 1976.

[166] A. A. Abrikosov, "On the magnetic properties of superconductors of the second group," *Sov. Phys. JETP*, vol. 5, pp. 1174–1182, 1957.

[167] D. Diakonov and V. Y. Petrov, "A formula for the Wilson loop," *Phys. Lett.*, vol. B224, pp. 131–135, 1989.

[168] D. Diakonov and V. Petrov, "Non-Abelian Stokes theorem and quark monopole interaction," 1996.

[169] K.-I. Kondo, "Abelian magnetic monopole dominance in quark confinement," *Phys. Rev.*, vol. D58, p. 105016, 1998.

[170] S. Weinberg, *The Quantum Theory of Fields: Volume II Modern Applications.* Cambridge, UK: Cambridge University Press, 2005.

[171] K.-I. Kondo, T. Murakami, and T. Shinohara, "BRST symmetry of SU(2) Yang-Mills theory in Cho-Faddeev-Niemi decomposition," *Eur. Phys. J.*, vol. C42, pp. 475–481, 2005.

[172] N. Weiss, "The effective potential for the order parameter of gauge theories at finite temperature," *Phys. Rev.*, vol. D24, p. 475, 1981.

[173] F. Marhauser and J. M. Pawlowski, "Confinement in Polyakov gauge," 2008.

[174] C. Wetterich, "Exact evolution equation for the effective potential," *Phys. Lett.*, vol. B301, pp. 90–94, 1993.

[175] J. Berges, N. Tetradis, and C. Wetterich, "Nonperturbative renormalization flow in quantum field theory and statistical physics," *Phys. Rept.*, vol. 363, pp. 223–386, 2002.

[176] J. M. Pawlowski, "Aspects of the functional renormalisation group," *Annals Phys.*, vol. 322, pp. 2831–2915, 2007.

[177] J. Braun, H. Gies, and J. M. Pawlowski, "Quark confinement from color confinement," *Phys. Lett.*, vol. B684, pp. 262–267, 2010.

[178] N. M. Bratovic, T. Hell, and W. Weise, *to be published.* 2010.

[179] T. Hatsuda, M. Tachibana, N. Yamamoto, and G. Baym, "New critical point induced by the axial anomaly in dense QCD," *Phys. Rev. Lett.*, vol. 97, p. 122001, 2006.

[180] N. Yamamoto, M. Tachibana, T. Hatsuda, and G. Baym, "Phase structure, collective modes, and the axial anomaly in dense QCD," *Phys. Rev.*, vol. D76, p. 074001, 2007.

Acknowledgments

In these final lines I would like to express my thanks and appreciation to those people who have supported me during the writing of this work and previously, during my studies in Munich.

First of all, I am indebted to Prof. Dr. Wolfram Weise who offered me the possibility to work out the nonlocal version of the Polyakov-loop-extended Nambu–Jona-Lasinio model, developed in part at T39 in the years before I joined the group. I would like to thank Wolfram Weise for many inspiring discussions, suggestions and comments concerning my work. Furthermore, I thank him for giving me the opportunity to participate in various workshops and conferences in the United States of America, Japan and Italy.

Speaking of financial support, I would like to mention the German Bundesministerium für Bildung und Forschung (BMBF), the Gesellschaft für Schwerionenforschung (GSI), and the Excellence Cluster "Origin and Structure of the Universe" of the Deutsche Forschungsgemeinschaft (DFG).

I thank Jean-Paul Blaizot and Georges Ripka at the ECT* in Trento for hosting me.

I am grateful to Volker Koch, Gunther Roland and Mikhail Stephanov who invited me to the INT workshop "The QCD Critical Point" at the Institute for Nuclear Theory (INT) in Seattle. In addition, I thank the INT for financial support and accommodation during my 30-days-stay in Seattle. Furthermore, I thank the Brookhaven National Laboratory (BNL) for hosting me during the "Critical Point and Onset of Deconfinement" conference in 2009.

I am mostly indebted to Kenji Fukushima who invited me to the Yukawa Institute for Theoretical Physics in Kyoto to attend the workshop "New Frontiers in QCD 2010—Exotic Hadron Systems and Dense Matter" in March 2010. I would like to thank all the organizers for a very instructive workshop, for many discussions and for many joint dinners. I thank the Yukawa Institute for accommodation and financial support.

Thanks go to the Graduate School of the Technische Universität München for financial support and for giving me the opportunity to enlarge my teaching and rhetoric skills. I thank the Graduate School for supporting financially my guest Motoi Tachibana during his research visit in Munich.

Finally, I would like to express my thanks to Norbert Kaiser with whom I had many discussions on physical and, in particular, mathematical questions arising during the writing of my work. I would also like to mention Simon Rößner, Marco Cristoforetti, Bertram Klein, Nino Bratovic and Kouji Kashiwa who have been working together with me on the PNJL model. I thank Bernhard Musch and Alexander Laschka for their advice with computer trouble. Alexander Laschka and Kouji Kashiwa and my former office mates Marco Cristoforetti, Youngshin Kwon, Massimiliano Procura and Marina Dorati have always contributed to a very relaxing and friendly atmosphere.

I thank my other colleagues from T39, Antonio Vairo, Philipp Hägler, Lisheng Geng, Jeremy Holt, Michael Altenbuchinger, Salvatore Fiorilla and Robert Lang for many discussions and for shared hours outside the offices.

Last, but not least, I thank my family for their support during the period of my studies in Munich. I apologize that these few lines are not enough to express my gratitude towards my family that it would actually deserve!

VSG

VDM Verlagsservicegesellschaft mbH

Die VDM Verlagsservicegesellschaft sucht für wissen-
schaftliche Verlage abgeschlossene und herausragende

Dissertationen, Habilitationen, Diplomarbeiten, Master Theses, Magisterarbeiten usw.

für die kostenlose Publikation als Fachbuch.

Sie verfügen über eine Arbeit, die hohen inhaltlichen und for-
malen Ansprüchen genügt, und haben Interesse an einer hono-
rarvergüteten Publikation?

Dann senden Sie bitte erste Informationen über sich und Ihre
Arbeit per Email an *info@vdm-vsg.de*.

Sie erhalten kurzfristig unser Feedback!

VDM Verlagsservicegesellschaft mbH
Dudweiler Landstr. 99
D - 66123 Saarbrücken

Telefon +49 681 3720 174
Fax   +49 681 3720 1749

www.vdm-vsg.de

Die VDM Verlagsservicegesellschaft mbH vertritt

VDM
Verlag
Dr. Müller

LAP LAMBERT
Academic Publishing

SVH Südwestdeutscher Verlag
für Hochschulschriften

www.ingramcontent.com/pod-product-compliance
Lightning Source LLC
Chambersburg PA
CBHW021056210326
41598CB00016B/1224